Smart and Intelligent Systems

The Human Element in Smart and Intelligent Systems

Series editor: Abbas Moallem

This series aims to cover the role of the human element in all aspects of smart and intelligent systems. It will include a broad range of reference works, textbooks, and handbooks. The series will look for single-authored works and edited collections that include, but not limited to: automated driving, smart networks and devices, cybersecurity, data visualization and analysis, social networking, smart cities, smart manufacturing, trust and privacy, artificial intelligence, cognitive intelligence, pattern recognition, computational intelligence, and robotics. Both introductory and advanced material for students and professionals will be included.

Smart and Intelligent Systems
The Human Elements in Artificial Intelligence, Robotics, and Cybersecurity
Abbas Moallem

For more information on this series, please visit: https://www.routledge.com/The-Human-Element-in-Smart-and-Intelligent-Systems/book-series/CRCHESIS

Smart and Intelligent Systems

The Human Elements in Artificial Intelligence, Robotics, and Cybersecurity

Edited by
Abbas Moallem

CRC Press
Taylor & Francis Group
Boca Raton London New York

CRC Press is an imprint of the
Taylor & Francis Group, an **informa** business

First edition published 2022
by CRC Press
6000 Broken Sound Parkway NW, Suite 300, Boca Raton, FL 33487-2742

and by CRC Press
2 Park Square, Milton Park, Abingdon, Oxon, OX14 4RN

CRC Press is an imprint of Taylor & Francis Group, LLC

© 2022 Taylor & Francis Group, LLC

ISBN: 978-0-367-46149-2 (hbk)
ISBN: 978-1-032-10440-9 (pbk)
ISBN: 978-1-003-21534-9 (ebk)

DOI: 10.1201/9781003215349

Typeset in Times
by SPi Technologies India Pvt Ltd (Straive)

Contents

Editor

Dr. Abbas Moallem is a consultant and adjunct professor at San Jose State University, California where he teaches human–computer interaction, cybersecurity, information visualization, and human factors. He is the program chair of HCI-CPT, the International Conference on HCI for Cybersecurity, Privacy and Trust.

Contributors

Dr. Mohd Anwar is a professor of Computer Science at North Carolina A&T State University. His research has contributed to various topics in cybersecurity including intrusion/malware detection, usable security, cyber identity, and differential privacy. He employs AI/ML methods to cybersecurity. He is particularly interested in trustworthiness and replicability of ML-based models. Dr. Anwar has more than 100 peer-reviewed publications, and his research has extensively been funded by NSF, DoD, Air Force, NSA, and NIH.

Dr. Jessie Y.C. Chen is a Senior Research Scientist for Soldier Performance in Socio-Technical Systems with U.S. Army Research Laboratory. Her research interests include human-autonomy teaming, human–robot interaction, and agent transparency. Dr. Chen is an Associate Editor for *IEEE Transactions on Human-Machine Systems* and *IEEE Robotics & Automation – Letters*.

Dr. Xiangfeng Dai is currently an Application Analyst at the University of North Carolina at Chapel Hill. Dr. Dai has an interdisciplinary background, and his research interests focus on machine learning, natural language processing, and health informatics. Dr. Dai published numerous papers in the fields and regularly serves as a peer reviewer for leading journals and serves as a chair for conferences. In addition, Dr. Dai has many years of industry experience and developed various cutting-edge data-driven applications.

Celso M. de Melo is a computer scientist at the CCDC Army Research Laboratory with an interest on artificial intelligence and human–machine interaction. His research focuses on using synthetic data to improve machine learning performance, develop general theory of human–machine collaboration, and design (socially) intelligent machines.

Dr. Benjamin T. Files researches how measurable individual variability across timescales can be understood and used to build personalized systems and interactions with information. His research techniques include computational modeling, behavioral studies, and physiological measures. Dr. Files has been a civilian scientist at the U.S. Army Research Laboratory since 2016.

Dr. Hilda Goins earned her doctoral, master's and bachelor's degrees from North Carolina Agricultural and Technical State University. She is currently a postdoctoral research associate at North Carolina Agricultural and Technical State University. Dr. Goins' background includes artificial intelligence and machine learning applications for daily life, image processing, multi-resolution, and wavelet analysis.

Dr. Peter Khooshabeh is the regional lead for the West Coast U.S. Army Combat Capabilities Development Command's Army Research Laboratory.

Carl Markert (carlmarkert@tamu.edu) is a Ph.D. student in the Department of Industrial and Systems Engineering at Texas A&M University. His research interests are focused on ways to enable seniors to live at home safely, independently, and comfortably and is concentrated on understanding the effectiveness of telehealth combined with health coaching via mobile devices.

Dr. Farhad Mehdipour is an academic and R&D expert with over 20 years of experience both in industry and academia. Farhad has initiated and led several interdisciplinary R&D projects and published 100+ peer-reviewed articles. Farhad is currently a head of department, and a principal lecturer at Otago Polytechnic – Auckland Campus, and an adjunct professor at St Bonaventure University, USA. He is a senior member of IEEE.

Dr. Mahnaz Moallem is a Professor of Learning Design and Research and Chair of the Department of Educational Technology and Literacy, College of Education, Towson University, Maryland. Dr. Moallem received her Ph.D. in Learning Systems Design and her Program Evaluation Certificate from Florida State University, USA. She has a broad background in learning sciences, with specific expertise in emerging technology integration, learning theories, and educational research. Her experience as a teacher, teacher educator, curriculum designer, and program evaluator inform her research.

Jukrin Moon (jukrin.moon@tamu.edu) is a Ph.D. student in the Department of Industrial and Systems Engineering at Texas A&M University. Her research interests center around understanding interactions among humans and other cognitive systems, thereby better designing and managing networked systems' adaptive behaviors in safety-critical domains

Dr. Kimberly A. Pollard is a biologist at the U.S. Army Research Laboratory, where she specializes in research on communication and perception, training, and human–agent interaction. Her work examines the ways in which information presentation, individual differences, task properties, and social elements come together to affect performance in human-technology domains.

Dr. Hossein Sarrafzadeh is a university distinguished professor and director of Center of Excellence in Cybersecurity at NC A&T State University. He has worked in data mining and machine learning in cybersecurity, IoT, and cybersecurity in Smart Cities. He developed one of the world's first real-time facial expression and gesture recognition systems for emotion recognition. He has founded several cybersecurity research and operations centers in NZ, Australia, and the USA. He has several inventions and patented systems in computer vision, cybersecurity, and cloud security. He has published over 180 research articles and secured over $15M funding in the past 5 years.

Dr. Farzan Sasangohar (sasangohar@tamu.edu) is an assistant professor of Industrial and Systems Engineering at Texas A&M University. He is also an assistant professor of Outcomes Research at Houston Methodist Hospital. Farzan received his Ph.D. in Industrial Engineering from University of Toronto and his SM in Engineering Systems from MIT.

Dr. Jessica Schwarz is leading a research group at Fraunhofer FKIE, Germany. Her expertise is focused on Human Factors research in (adaptive) human–machine interaction. She is co-chairing the international conference "Adaptive Instructional Systems" (affiliated to HCII) and involved in NATO research activities related to measuring the cognitive load on the soldier.

Tianyang Zhang received the B.Sc. degree in electrical engineering from North Carolina A&T State University, NC, USA, and Henan Polytechnic University, Henan, China, in 2017, and the M.Sc. degree in electrical and computer engineering from the University of Massachusetts Amherst, MA, USA, in 2019. He is currently a machine learning engineer in Learnable AI. He has worked as a Research Associate at the Department of Computer Science, North Carolina A&T State University, NC.

Introduction

In recent history, computers have changed our view of systems design. Mass data, machine learning, artificial intelligence (AI), and the Internet of Things (IoT) have further optimized and revolutionized the system's design. The new and immersive technologies transformed our understanding of systems architecture and design. They brought along a new framework to describe systems that are often labeled as "intelligent" or "smart."

BUT WHAT EXACTLY MAKES A SYSTEM SMART OR INTELLIGENT?

"Smart" and "intelligent" are words that are a part of our everyday language. At first glance, they might seem easily interchangeable. However, there is a difference between what we mean and how we use these words. In other words, the distinction is reflected in the context in which the terms are used. For example, when one refers to a friend as "intelligent," they do not necessarily mean that the friend is also "smart." When the term "intelligent" is used most of the time, it means someone has the brain-processing power to solve problems, measured by evaluating the person's reasoning ability. When the term "smart" is used, it often means using intelligence effectively and practically in day-to-day life. The same might also be applied to computer systems. A computer system may be considered intelligent when it has strong processing power but will not be capable of doing much to be considered smart without any applications. Having processing power without the ability to use it makes the difference between smart and intelligent systems.

With advances in mass data, machine learning, and artificial intelligence, from education to manufacturing and military, to name a few, computer systems have become more and more intelligent and smart. However, human agents continue to play an essential role in decision-making and monitoring the automated system despite all these advances. Consequently, analysis of the human factors in smart and intelligent systems is a crucial part of the system design. The agents need to understand the functional aspects of the tasks and the broader social context. As more systems become automated and more sophisticated, the nature of human–agent interaction is also profoundly changing. One of the pioneers of human–machine interfaces, Licklider[1], called this concept "Man-Computer Symbiosis." He believes that automation had given way to the replacement of men to automation, and "the men who remain are there more to help than to be helped." According to Licklider, in large computer-centered information and controlled systems, the human operators are responsible mainly for functions that prove infeasible to automate.

The market for artificial intelligence (AI) systems is also growing. According to Statista[2], worldwide market revenues from the cognitive and AI systems reached 50.1 billion U.S. dollars. By the end of 2018, there were an estimated 22 billion IoT-connected devices in use around the world[3].

The smart and intelligent system's usage expanded to all industries, from manufacturing, education, healthcare to military and defense systems. The consumers' smart, intelligent systems also grew. The *Verge*[4] reported in 2019 that more than 100 million devices with Alexa on board have been sold. North American consumers are expected to spend the most on smart home systems and services in 2022, and the number of smart homes is forecast to grow and surpass the 300 million mark by 2023[5].

With this expansion in mind, studying and research in intelligent systems, particularly interaction with humans, become significant and face numerous challenges. We need to understand and gain knowledge from various aspects, including programming, human behavior, machine learning, artificial intelligence, and cybersecurity. This book's objective is to provide snapshots into a selected number of topics (considering the extent of this field) related to designing smart and intelligent systems. The following paragraphs provide a summary of the areas addressed in the book.

One of the areas that significantly impacted by smart technologies is the educational system. Gradually, smart technologies transformed traditional education by enabling, for example, learners to access more effective educational resources. Additionally, a smart educational system has instigated new areas of educational research and prompted a new theoretical understanding of foundational knowledge and skills. Thus, in the first two chapters, the authors explore how smart technologies are transforming educational systems. In Chapter 1, the author offers an extensive review of smart educational technologies, their advantages, and the challenges facing learners and educators. Chapter 2's authors introduce a tutoring system that utilizes an existing generic affective application model to infer and respond to the learner's affective state.

Chapter 3, entitled "User state assessment in adaptive system design," investigates how the smart and intelligent system helps human operators augment their cognitions and facilitate decision-making. The author acknowledge that the new system design adds a tremendous workload on human operators and further discuss how multi-dimensional state assessment such as fatigue, situation awareness, emotional state, and environmental and context of actions influence the system design. The technological progress, specifically in sensor technology, opens various possibilities to detect mental states in adaptive technology user state assessment. The author expand our understanding of the methodologies derived from evaluating human–computer interaction systems and information retrieval and information filtering systems. They offer several examples of how applying these methodologies in user-adaptive systems can be conducted.

Chapter 4, entitled "Agent transparency," introduces the readers to the machine and human agents and more autonomous intelligent agents and explains how humans are increasingly working with agents. The author argue that robots are gradually becoming human collaborators rather than a tool that human uses. To make human and agents' collaboration and teaming successful, agent transparency is critical. Chapter 4 further shares the research findings and application on agent transparency that includes operator trust calibration, situation awareness, workload, and individual and cultural differences. The author conclude these elements need to be applied in user interface designs in smart and intelligent systems.

Chapter 5, entitled "Smart telehealth systems for the aging population" discusses that one of the extensively profited areas from machine learning and artificial intelligence in smart and intelligent systems is the healthcare industry. The authors argue that the healthcare systems have undergone an extensive transformation from wearable monitoring devices, measurement equipment, augmented cognition devices to the clinical decision support system, or even robots performing healthcare tasks. The pandemic of 2020 accelerated these changes like education systems, from visiting virtually the doctors to the administrative control of patients and monitoring the pandemics is changing rapidly. The smart telehealth systems for the aging population are a good example of this promising transformative system for patients and healthcare providers. In telehealth systems, various technologies are used. The patients are remotely monitored, the physiological and biological parameters are captured, decision support systems are used to analyze the data, and health-related behavior changes are monitored through health coaching systems.

Chapter 6, entitled "Social factors in human-agent teaming" describes another challenge related to progress in developing autonomous technology, such as robots, drones, self-driving cars, and personal assistants. It focuses on the sophistication of the interaction with the surrounding environment. The authors argue that to make this interaction and collaboration with human agents effective and efficient, the agents need to understand the task's functional aspects and the broader social context. It is suggested that considering psychological theories, how and when humans treat agents in a social manner is another dimension in designing smart and intelligent systems. The analysis of the experimental evidence from natural language conversation to emotional expressions communication is another dimension to make collaboration between humans and agents successful.

"Human elements in machine learning-based solutions to cybersecurity" is the title of Chapter 7. The chapter's authors argue that one of the main components of the smart and intelligent system's success is ML. They further explain that ML algorithms enabled smart systems to become smarter in various areas such as cybersecurity, phishing detection, malware detection, and intrusion detection. Machine learning in business helps in enhancing business scalability and improving business operations for companies across the globe. Artificial intelligence tools and numerous ML algorithms have gained tremendous popularity in the business analytics community. Interactive machine learning provides feedback to human agents in response to the model's output.

Finally, Chapter 8, entitled the "Cybersecurity in smart and intelligent manufacturing systems" reminds the reader that with growing cyberattacks, cybersecurity has become an essential area in designing smart and intelligent systems in recent decades. The risks of cyberattacks are exponentially increasing every day, adding substantial financial impacts on enterprises. One of the areas that went through a transformation with the smart and intelligent systems is manufacturing systems with the interconnection of smart machinery and IoT devices within a network of computers in a closed or open networking design. Chapter 8 provides an overview of cybersecurity in smart and intelligent manufacturing systems by discussing the motivation of cyberattackers, the manufacturing of cyber vulnerabilities, and protective technologies in smart and intelligent manufacturing systems.

In summary, with growing research and applications in smart and intelligent systems, I hope this book will provide academics, researchers, practitioners, and students an insight on the selected topics in smart and intelligent systems. The book was put together during the unprecedented period of the 2020 Covid pandemic. I am grateful to all contributors for their commitment and contributions despite the hardship of this time.

Abbas Moallem

NOTES

1 Licklider J.C.R. (1960): "Man-computer symbiosis," *IRE Transactions on Human Factors in Electronics*, volume HFE-1, pages 4–11, March 1960, https://groups.csail.mit.edu/medg/people/psz/Licklider.html
2 https://www.statista.com/statistics/694638/worldwide-cognitive-and-artificial-intelligence-revenues/
3 https://www.statista.com/statistics/802690/worldwide-connected-devices-by-access-technology/
4 Dieter Bohn (2019): "Amazon says 100 million Alexa devices have been sold—what's next?," *The Verge*, Jan 4, 2019. https://www.theverge.com/2019/1/4/18168565/amazon-alexa-devices-how-many-sold-number-100-million-dave-limp
5 https://www.statista.com/topics/2430/smart-homes/

1 Smart Educational System

Mahnaz Moallem

CONTENTS

DOI: 10.1201/9781003215349-1

1.1 INTRODUCTION

Information and Communication Technology (I.C.T.) has revolutionized our lives and how we interact with each other. We now live, work, and learn in a world surrounded by multiple interconnected devices (computers, tablets, mobile phones, cameras, and robots). I.C.T. has also provided us with an opportunity to transform the education system to a smarter system in which there is not a singular technology connecting us, but it is a blend of various hardware (smart devices) and software intelligent technologies [e.g., network, cloud computing, sensors, big data or learning analytics, and artificial intelligence (A.I.)] enabling communication anytime, anywhere, and among anything. Furthermore, the technologically smarter education system depends on a new computing architecture or infrastructure that expands the human-to-human interaction. It interconnects everyday items [The Internet of Things (IoT)] in our lives, making the Internet even more immersive and ubiquitous. Such a technologically sophisticated smart system creates unique and exceptional opportunities for the education system to use intelligent technologies to capture, analyze, direct, and improve learning and teaching and support the development of flexible, adaptive, and personalized learning (Mayer-Schönberger & Cukier, 2013; Picciano, 2012).

Using an advanced, immersive, and ubiquitous computing infrastructure is not the only component of the smart education system. Today, there is a noticeable shift in pedagogical approaches that constitute smart education. In other words, smart education represents creating an innovative learning environment that is essentially different from the so-called traditional learning environments. The new approaches for learning and the shift in instructional paradigm embrace a learning environment that demonstrates significant maturity at various "smartness" levels or smart features such as adaptation, flexibility, contextual awareness, prediction (analytics/advanced learner profiling), inferences (big data), self-learning, self-organization, and collaboration (Zhu, Sun, & Riezebos, 2016).

But how can we define the concept of smart education and smart learning, given their many different conceptualizations and features? How can we explain smart education related to other concepts such as IoT, augmented, or virtual reality (V.R.), blockchain innovation? What does flexible, adaptive, and personalized learning represent in smart education? How does the smart aspect come into play in the context of A.I., machine or deep learning, big data, or similar concepts? How such a smart system is developed, and how is it tailored to the education system and learners' needs and socio-cultural context?

The purpose of this chapter is to explore the concepts of smart education, smart learning, and smart learning environment (S.L.E.), and the above-listed questions to explain what they mean. It will further examine how smart technologies and shifts in learning and instructional approaches disrupt traditional education and move it to a smarter learning environment. It also discusses how technological development and the establishment of smart learning have resulted in new foundational knowledge and skills, theoretical frameworks, and field of research.

1.2 THE EVOLUTION OF SMART EDUCATION AND SMART LEARNING

With the development of advanced technologies, and their impact on learning environments, the concepts of smart education and smart learning emerged. However, there is still not a clear and unified definition of smart education or smart learning. Besides, the two concepts of smart education and smart learning have been used interchangeably. Thus, it is not clear how the two concepts are similar or different.

1.3 SMART EDUCATION

Some scholars refer to smart education as a paradigm change in the way students access education. They suggest that smart education is the educational system that allows students to learn by using up-to-date technologies (e.g., Gros, 2016a). It enables them to learn anytime and anywhere through the technologies offered in the smart learning environment. Others emphasize that the concept of smart education describes learning in the digital age and goes beyond a change in the delivery of education and refers to a radical change in the ways teaching and learning occur (e.g., Uskov, Bakken, & Aluri, 2019). Consistent with this view, Zhu, Yu, et al. (2016) define smart education as "intelligent, personalized, and adaptive education" (page 2). Using smart technologies, they argue that smart pedagogies can be adopted to provide personalized learning and empower learners to develop higher thinking quality (Zhu, Yu, et al., 2016). They further introduce a smart education framework, identifying three core elements of teacher presence, learner presence, and technology presence (Hoel & Mason, 2018; Zhu, Yu, et al., 2016). The teaching presence describes the teacher's role in a smart education system as a designer of a student-centered, personalized, and collaborative learning environment, facilitating the learning process by promoting interactions, providing feedback/direct instructions, and supporting learners to use technologies. The learner presence refers to autonomous and collaborative learners who are efficient users of technologies. The technological presence describes advanced technologies that provide ubiquitous access to learning resources and adapt to learners' individual needs (Zhu, Yu, et al., 2016). According to this framework, smart education bases its foundations on smart pedagogies, smart devices, and intelligent technologies. Accordingly, Horizon 2020 report refers to the following list of six emerging technologies and practices that make intelligent, personalized, and adaptive education possible.

- "Adaptive Learning Technologies
- A.I./Machine Learning Education Applications
- Analytics for Student Success
- Elevation of Instructional Design, Learning Engineering, and UX Design in Pedagogy
- Open Educational Resources
- X.R. (A.R./VR/MR/Haptic) Technologies" (2020 Educause Horizon Report, page 5)

The Annenberg Institute for School Reform (AISR) at Brown University (Voices in Urban Education, 2010) offers a much broader definition of smart education. It describes that a smart education system links a high-functioning school district with a web of supports. Such systems require collaboration between the school district, city agencies, cultural institutions, community groups, and businesses. It points to four key areas that support the principles of the smart education system: capacity building in school districts, capacity building in communities, research and knowledge product development, communication, dissemination, and learning opportunities.

The synthesis of the current literature on the smart education system suggests that the core element that differentiates this system from traditional education is smart and intelligent technologies (both hardware and software). Four essential elements further explain the core elements of a smart education system (see Figure 1.1). To deploy a smart education system, institutions (school districts, higher education institutions, and corporate training) need to build *an advanced, immersive, and ubiquitous computing/technology infrastructure.* An advanced, immersive, and ubiquitous computing/technology infrastructure provides a foundation to access digital, context-aware, and adaptive devices, web technology, sensors tracking, A.I., machine learning, deep learning, big data, complex interface engines, virtual, augmented mixed reality and Haptic technologies, and analytics for student success. Furthermore, smart and intelligent technologies with advanced infrastructure make it possible to create technology-facilitated learning or *smart learning environment* in which new and innovative learning approaches such as flexible learning (learning anywhere, anytime), personalized and adaptive learning, integrated formal and informal learning, student-centered, interactive, and self-directed learning approaches become a reality. The smart education system does not work independently. It operates in a *collaborative environment.* In a smart education system, school districts or universities collaborate with community groups, cultural institutions, businesses, and city agencies to provide open, flexible, personalized, and adaptive education. A smart

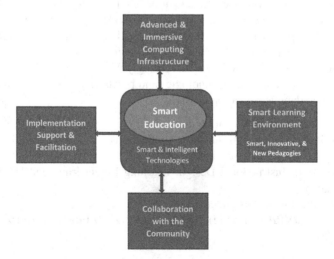

FIGURE 1.1   Components of smart education system.

education system, moreover, cannot operate without robust *implementation support and facilitation.* Technology, instructional design, universal design, and teacher's support are needed to provide for students' success in the smart education system (see Figure 1.1).

1.3.1 ADVANCED AND IMMERSIVE COMPUTING INFRASTRUCTURE

Smart education revolves around advanced and immersive computing infrastructure. Similar to the concept of a smart city (Nikolov et al., 2016), smart education systems are those in which computing infrastructure makes it possible for information technology (I.T.) solutions (e.g., A.I., IoT, analytics, blockchain, X.R., machine and machine deep learning, big data, etc.) to be deployed to make the operation of smart learning environments viable and sustainable. Outdated systems, networks, hardware and software, and processes are unable to support innovative, intelligent computing that strives to provide flexible, adaptive, and personalized learning and expand student learning to anywhere, anytime, anyplace, and anything. Smart education systems must have a multilayer computing infrastructure to enable smart learning environments to offer advanced, intelligent, immersive I.T. solutions. The first layer of computing infrastructure consists of the following (see Figure 1.2).

- *High-speed broadband Internet infrastructure.* A smart education system needs an Internet infrastructure that maintains an up-to-date and sophisticated Wi-Fi infrastructure plan that includes smart strategies and solutions to prepare the campuses/schools/communities for what can come in the future.
- *Network security infrastructure.* A smart education system must have a structure that ensures network security to protect sensitive or personal information.

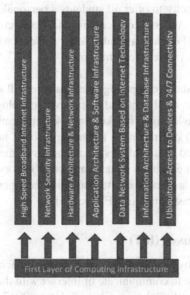

FIGURE 1.2   Advanced and immersive computing infrastructure.

The rigorous security measures must be built into the network designs. Cloud-based technologies and virtualization are needed to cope with the increased volume of data and devices. With cloud computing, security management architecture integrates hardware, such as server equipment, storage devices, and network equipment into a vast hardware device virtualization resource pool.

- *Hardware architecture and network infrastructure.* In an advanced smart education system, various hardware device resources (e.g., large-scale server equipment, storage devices, and network equipment) are integrated to reduce the space occupied by the hardware devices to fully utilize resources.
- *Application architecture and software infrastructure.* A smart education system uses a cloud computing network that offers users storage services through a non-local or remote distributed computer for various software systems, such as student management information system software, office management system software, teaching management system software, and experimental management system software.
- *Data network system based on Internet technology.* A smart education system has network infrastructures that collect data from interconnected mechanical and digital devices, objects, animals, people with unique identifiers (IoT). It transfers the data without requiring human to human or human to computer interaction. These networks are 4G LTE and 5G built to support the IoT resource demand.
- *Information architecture and database infrastructure.* A smart education system needs to create an ecosystem that fosters personalized, adaptive learning, community, and connectedness. This ecosystem includes robust, scalable, and agile architecture and infrastructures, including data network systems on Internet technology.
- *Ubiquitous access to devices and 24/7 connectivity.* Smart learning systems must have the capacity to offer all students a personal, connected device anytime and anyplace. Access to devices and 24/7 connectivity will address equity by making personal devices available to all students wherever they need them and Internet access in every learning space.

The advanced and updated computing infrastructure will support implementing a second layer of innovative, intelligent, and immersive technologies, platforms, and processes. IoT, A.I., machine learning, blockchain-based application, virtual, augmented, and mixed reality (X.R.) or immersive technology, social networking/collaborating technologies, big data, and data analytics are examples of the second layer emerging technologies across smart education systems and other sectors and organizations such as health and training industries. These technologies can operate on the first layer of advanced technology infrastructure (see Figure 1.3).

1.3.2   INTERNET OF THINGS (IoT)

IoT refers to a "global physical network that connects devices, objects and things to the Internet infrastructure to communicate or interact with the internal and the external environment" (Aldowah, Rehman, Ghaz, & Umar, 2017, p. 2). It is a system of

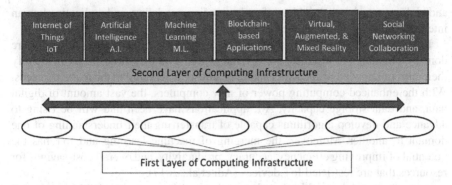

FIGURE 1.3 The second layer of computing infrastructure.

sensors or devices that collects and exchanges data through a network connection without any human involvement. IoT can monitor the state of a machine, the environment, or even a person, the "thing" in the IoT. IoT devices are non-standard computing devices (e.g., interactive displays, cloud-based learning materials, interactive video technology, scan maker, e-learning, and mobile real-time tools such as games, audio/video animation, Learning Management systems, etc.) that connect wirelessly to a network. They can transmit data. IoT will enhance the student learning experiences by creating opportunities and seamless access to instructional content in structured and unstructured formats (Aldowah et al., 2017). Furthermore, using IoT, learning experiences will continue to become more virtual, and classrooms will be better equipped for learning.

IoT devices can generate massive amounts of data useful for machine learning, predictive, and analytics. IoT data requires the infrastructure to manage better, store, and analyze a large volume of structured and unstructured data in real time (Adi, Anwar, Baig, & Zeadally, 2020). Given the enormous volume of data that is locally handled and transferred to be further processed or analyzed to produce knowledge, the role of data analytics for IoT data processing will be crucial (Adi et al., 2017). Machine learning (a group of techniques for analyzing data with no human intervention) will facilitate the quick processing of large-volume data emerging from IoT devices.

1.3.3 ARTIFICIAL INTELLIGENCE AND MACHINE LEARNING

A.I. refers to the computing technologies (algorithms or rule-based systems using if–then conditions) that resemble processes associated with human intelligence, such as reasoning, deep learning, adaptation, sensory understanding, and interaction (Nuffield Council on Bioethics, 2018; Tran et al., 2019). A.I. technologies are capable of learning from historical data, which makes them usable for an array of solutions such as robotics, self-driving cars, power grid optimization, and natural language understanding (Petersson, 2020; Stanford University, 2016). From a broader perspective, A.I. is defined as an interdisciplinary approach that adopts principles and devices from a variety of fields, such as computation, mathematics, logic, and biology, to solve the problem of understanding, modeling, and replicating human intelligence and cognitive processes (Tran et al., 2019).

In education, A.I. will direct computers to perform tasks that traditionally are done by a human. For instance, through interaction with the learner, A.I. can identify their current knowledge and skills and then adjust the learning to the learner's needs. With the enhanced computing power of new computers, the vast amount of digital data, and data storage capacity, A.I. applications have been and will be rising to advance and develop algorithms capable of transferring their understanding of one domain to another. Moreover, the merging of machine learning and IoT has the potential of improving efficiency, accuracy, productivity, and overall cost-savings for resources that are restricted IoT devices (Adi et al., 2017).

As a subset of A.I., machine learning, or deep machine learning (D.M.L.), is one of the A.I. algorithms developed to mimic human intelligence that can transfer understanding from one domain to another. Adi et al. (2017) define machine learning as a "family of techniques for analyzing data, wherein the process of model building on training data is automated (i.e., requires little to no human intervention)," p. 16206. Before machine learning, computer scientists had to teach computers all the ins and outs of every decision they had to make through rule-based algorithms. Machine learning takes an entirely different approach and lets the machines learn by ingesting vast amounts of data and detecting patterns. Machine learning algorithms use statistical formulas and big data to function. The idea is that advancements in big data and the large volume of data we collect enable machine learning to detect patterns (Petersson, 2020).

D.M.L. was developed on biological neural networks enabling smarter results than possible with machine learning. In other words, D.M.L. uses layers of information processing, each gradually learning a more complex representation of data. Advances in D.M.L. offer the ability to successfully train convolutional neural networks (extraction of common patterns found within local regions of the input images) with applications such as object recognition, video labeling, activity recognition, and many others. In education, D.M.L. has been used to predict dropouts and withdrawals in an on-going course and analyze the intrinsic factors impacting student performance. It has also been utilized for student assessment to predict students at risk of failure in the 2nd, 4th, and 9th week of their first year of engineering (Coelho & Silveira, 2017; Waheed et al., 2020). Researchers have begun showing the deep learning model's effectiveness in the early prediction of student performance, enabling timely intervention to implement corrective strategies for student support and counseling (e.g., Waheed et al., 2020). Interest in deep learning has increased substantially as it has made significant progress across a range of complex tasks (e.g., voice and image recognition) (Tenzin, Lemay, Basnet, & Bazelais, 2020).

1.3.4 BLOCKCHAIN-BASED APPLICATIONS

Blockchain is an emerging technology that started being widely implemented during 2019. It was first used as a peer-to-peer ledger for registering the transactions of Bitcoin cryptocurrency. The purpose was to reduce third-party dependency and allow users to make their transactions directly. Thus, it was designed as a decentralized peer-node network (Alammary, Alhazmi, Almasri, & Gillani, 2019). Blockchain was developed in three stages (1.0, 2.0, and 3.0). In Blockchain 3.0, many applications were developed in various sectors, including education. Although blockchain applications for education are still evolving, educational institutions have started to utilize blockchain technology. The current primary use of blockchain applications in education is to validate and share academic certificates or competencies and learning outcomes that students achieved. For instance, in some institutions, instead of relying on paper diplomas or alterable PDF files and lengthy requests for verification, students present employers with digital credentials that would be immediately and securely verified. The advancement in blockchain applications in education will improve data protection, privacy, and integrity, providing better control on how students' data are accessed and by whom, enhancing accountability and transparency, improving trust, and lowering the cost associated with transactions and storage of data. Blockchain technology can also enhance the assessment of students' performance and achievement of learning outcomes. Nonetheless, it is anticipated that blockchain technology will have much more to offer and revolutionize education (Alammary et al., 2019).

1.3.5 VIRTUAL, AUGMENTED, MIXED REALITY (X.R.), OR IMMERSIVE TECHNOLOGY

V.R. is a computer simulation that displays an environment through which the learner can walk and interact with simulated people (often called "agents" or "avatars"). The virtual environment is depicted as a three-dimensional world. It often tries to replicate "the real world both in appearance and in the way that objects behave (e.g., the simulation of gravity)" (Christou, 2010, p. 4). However, virtual space does not have to be like the real world. The virtual environments can be used to depict entirely unrealistic scenarios for training purposes and simulate the environment in which the learner will eventually operate by providing a safe environment to test scenarios that otherwise will be too difficult or dangerous to perform in real life (Christou, 2010). Thus, V.R. as a medium is "unreal" and relies on *perceptual stimulation* (including perceptual feedback on one's own action—visual cues, sounds, and sometimes touch and smell to trigger emotional reactions). Another V.R. phenomenon linked to emotional experience is *presence*. Presence is a dimensional construct and describes the extent to which a user feels present in a V.R. environment (Botella, Garcia-Palacios, Baños, & Quero, 2009; Schubert, Friedmann, & Regenbrecht, 2001; Slater & Wilbur, 1997). Researchers found three dimensions of presence: spatial presence, involvement, and realness (Schubert et al., 2001). Evidence in the literature indicates that the level of immersion a V.R. system exerts has an effect on the presence experienced by the user (Freeman, Lessiter, Pugh, & Keogh, 2005; Ijsselsteijn et al., 2001). Additionally, correlations between presence and emotional experience in V.R. have also been reported, especially in V.R. exposure therapy literature.

The availability of more advanced V.R. tools has made it possible for neuroscientists, psychologists, biologists, and other researchers to conduct more complex laboratory experiments on V.R.'s impacts on treatment and learning. For instance, V.R. has been used to assess its effects on new ways of applying psychological treatment or training (e.g., cerebral palsy, autism, fetal alcohol syndrome, and attention deficit) or on affective outcomes (e.g., anxiety, phobias) (Parsons, Rizzo, Rogers, & York, 2009;

Turner & Casey, 2014). Although still limited, V.R.'s effects in enhancing learning outcomes (e.g., motivation, presence, engagement, reflective thinking) have also been studied (e.g., Lee, Wong, & Fung, 2010; Makransky, Terkildsen, & Mayer, 2019). Simulations in V.R. will offer safe environments to practice and make mistakes without negative consequences. Learners can perform tasks, manipulate objects, explore environments, and do hands-on experience while being removed from anything that can cause harm to them.

Mixed or Augmented Reality (M.R. & A.R.) combines real and virtual worlds, supplementing the real world with computer-generated virtual objects in real time (Orman, Price, & Russell, 2017). Whereas V.R. completely immerses a user in a world that is entirely fictitious, A.R. overlays digital imagery over the existing physical world, "augmenting" it (Milgram & Kishino, 1994) as cited in Furht, 2011). A.R. is listed as a technology that has three essential requirements: combining real and virtual objects in a natural environment, aligning of real and virtual objects with each other, and real-time interaction. As such, a reflector or holographic sight would not be considered A.R. as, while it is a combination of digital and physical worlds, it is neither interactive nor does it blend in three dimensions.

A.R. provides new ways of interacting with the real world and can create experiences that otherwise would not be possible in either a completely real or virtual world. A.R. can facilitate the manipulation of virtual objects and observe phenomena that are difficult to observe in the real world (Lozano, 2018). Furthermore, A.R. can increase perceptual understanding of the phenomenon that is either invisible or difficult to observe as well as correct any misconceptions.

A.R. has been used more widely in education, both K-12 and higher education, compared with V.R., A.R. has been used in classrooms for learning subjects such as chemistry, mathematics, biology, physics, and astronomy. For instance, A.R. has shown to have increased content understanding (e.g., providing visual representations of the materials), memory retention, and motivation. A.R. has also been used outside of the classrooms (e.g., museum, zoo) and e-learning environments. Similarly, A.R. has been used for professional training across industries (e.g., space industry, military, manufacturing, etc.) and medical professionals through interactive visual representations, simulations, and the practice of various procedures. With the increased use of mobile computing devices (smartphones, cell phones, tablets), the potential of merging smartphones and A.R. for education is enormous. A.R. could provide students extra digital information about any objects and make complex information easier to understand.

In sum, the integrated real and virtual presentation and interaction technology have the potential of developing more powerful and supportive learning in smart learning environments (Andujar, Mejias, & Marquez, 2011; Kamarainen et al., 2013).

1.3.6 SOCIAL NETWORKING/COLLABORATING TECHNOLOGIES

Social networking platforms can be defined as a collection of networking, communication, sharing, and collaboration technology tools (asynchronous and synchronous) on Web 2.0 and Web 3.0. Examples of social networking are Facebook, Twitter, and Google. Examples of communication tools are discussion forums, videoconferencing tools, and chats, and examples of collaboration tools are blogs and wikis. Users of social networking sites share personal information through their profiles, connect with users of other sites, share multimedia content, link others to various web-accessible content, and initiate or join sub-groups based on shared interests or aspirations (Locker & Patterson, 2008). Social networking sites expand learning environments from classroom to formal and informal learning that can occur within and outside schools and institutional settings. Social networking offers the opportunity to develop a broader ecology of learning that facilitates a wide variety of activities and experiences of formal and informal learning, creating an ideal personalized learning environment (Khaddage, Müller, & Flintoff, 2016).

Collaborative technology tools "support the communication, coordination, and information processing needs of two or more people working together on a common task" (Galletta & Zhang, 2014, p. 145). Collaboration can also happen with robots and agents in addition to or instead of humans. Many technologies (synchronous and asynchronous) can be included in the category of collaborative technologies. However, the affordances of individual collaborative technologies are essential. According to Jeong and Hmelo-Silver (2012, 2016), computer-supported collaborative learning should afford learners opportunities to engage in joint tasks, communicate, share resources, participate in a productive collaborative process, and engage in co-construction of knowledge, monitor, and regulate collaborative learning and find and build groups and communities. Therefore, supporting collaborative learning with technologies means understanding the complexity of collaborative learning, along with the available pedagogies and technological tools (Jeong & Hmelo-Silver, 2016).

1.3.7 Big Data and Data Analytics

The term "Big Data" refers to any set of data generated from an increasing variety of sources, including Internet clicks, videos viewed, social media feeds, mobile phones, user-generated content, as well as purposefully generated content through sensor networks and other data generated in education or by educational institutions to manage courses, classes, and students. Big data requires powerful computational techniques to show trends and patterns within and between these very large data sets. Big data also refers to the tools and technologies used to store, analyze, and report data (data analytics). The challenges faced in processing big data technologies are overcome by using various statistical and data mining/analytics techniques (e.g., regression, nearest neighbor, clustering, classification). In their review of the literature on big data, Sin and Muthu (2015) found that big data techniques were used in a variety of ways in *learning analytics*, such as performance prediction (analyzing student's interaction in a learning environment with other students and teachers), attrition risk detection (analyzing student's behaviors to detect the risk of students dropping out from courses), intelligent feedback (intelligent and immediate feedback to students in response to their inputs to improve interaction and performance), course or learning materials recommendation (analyzing learner activities to identify their interests to recommend new courses or learning materials), student skill estimation (analyzing student information to estimate the skills acquired by the student), and grouping and collaboration of students, among several others.

Related to data analytics is the concept of *learning analytics*. Learning analytics is defined as specific analysis that focuses on collecting, analyzing, and reporting students' data to understand and improve their learning experiences to an optimum level (SoLAR, 2011; Siemens & Gasevic, 2012). In schools and other academic institutions, learning analytics serve as a foundation for personalized learning. By understanding and applying intelligent data, learner-produced data, and analysis models, learners are provided with personalized recommendations for appropriate learning resources, learning paths, or peer students through a recommending system (Maseleno et al., 2018).

1.3.8 SMART LEARNING ENVIRONMENT

The S.L.E. centers on learners and the content and context of learning more than on devices (Gwak, 2010; Shute et al., 2016). It is an intelligent environment tailored learning based on advanced I.C.T. infrastructure that provides adaptive, flexible, and personalized learning that supports and facilitates the learning operation (Gwak, 2010; Miraoui, 2018; Shute et al., 2016). A smart learning environment is not restricted only to formal learning activities but encompasses informal opportunities to provide an autonomous, adaptive, personalized learning environment (Kinshuk, Chen, Cheng, & Chew, 2016). According to Spector (2014), an S.L.E. is effective, efficient, and scalable; desirably engaging, flexible, adaptive, and personalized; and potentially is conversational, reflective, and innovative. Thus, the smart learning environment not only enables learners to access advanced technologies and digital resources and interact with learning systems anyplace, anytime, but it also provides learners with opportunities to personalize and self-direct their learning while interacting with a smart learning system capable of offering necessary learning guidance in real-world situations. The goal of a smart learning environment, therefore, is to provide self-learning, self-motivated, and personalized learning opportunities (e.g., context awareness, adaptive content, collaborative and interactive tool, rapid evaluation, and real-time feedback, etc.) for learners and allow them to access personalized learning content according to their differences (Kim, Cho, & Lee, 2013). Context-awareness or context-aware computing is one of the major features of the development of a smart learning environment. It refers to creating a context that provides task-relevant information to the learner based on the learner's learning behavior, learning process, and learning outcome. It combines a physical classroom with many virtual learning environments such as IoT, sensing, and other smart and wearable devices (e.g., brainwave detection, emotion recognition) (Kinshuk et al., 2016).

In sum, a smart learning environment is a technology-supported learning environment that uses an intelligent tutoring/support system to make adaptations and to provide support (guidance, feedback, hints, or tools) according to learners' needs, where needs are determined via analysis of learning and performance data in the online or real-world contexts (Hwang, 2014). Furthermore, the smart learning environment helps learners gain knowledge in formal and informal learning settings through context-aware learning systems [learning systems that detect real-world locations and contexts (location awareness, situation awareness, social awareness)]. It will allow the teacher/facilitator to focus more on teaching and keeping learners involved, and making learning deeper, more immediate, and more flexible. A smart learning environment calls for innovative or *smart pedagogical approaches* to orchestrate formal and informal learning. Thus, embedded in the smart learning environment are smart pedagogies that center on the following approaches.

1.3.9 SMART PEDAGOGICAL APPROACHES

1.3.9.1 Deep Learning Pedagogy

Deep learning pedagogy refers to learning that prioritizes problem-solving ability, adaptability, critical thinking, and the development of interpersonal, intrapersonal, and collaborative skills over rote memorization and the passive transmission of knowledge (Darling-Hammond & Oakes, 2019). According to Darling-Hammond and Oakes (2019), in learning environments, deep learning is emphasized, and challenging academic content is paired with engaging and innovative learning experiences. Such experiences equip students with skills to find, analyze, and apply knowledge in the new and emerging context. In other words, learning is contextualized, inquiry based, applied in real-world problems/tasks, and explicitly addresses issues of equity and social justice. Furthermore, deep learning problems or tasks are challenging and incorporate both curricular content and students' interests or aspirations, and through feedback and formative evaluation, learners' self-confidence, proactive dispositions, and self-reflection are improved.

The fundamental promise of deep learning pedagogy is that technology can play an indispensable role in assisting teachers in deepening and accelerating student learning across all subject areas. Using deep learning approaches, students can choose strategies based on their personal characteristics and the learning context. Students may modify their learning strategies depending on the learning context. With the support of advanced and intelligent technology, the teacher can support and guide students in achieving deep learning through critical analysis of the subject matter, questioning, guiding, and challenging assumptions.

Intelligent tutoring systems or pedagogical agents are examples of advanced and intelligent computer-supported technologies that assist in promoting deep learning and enhancing motivation. In a series of studies, Kim and Bayler (2016) simulated human instructional roles in anthropomorphic pedagogical agents. Grounding their work in social-cognitive perspectives of learning (affective experiences of learners influence their cognitive engagement leading to deeper learning), they developed three instructional roles for pedagogical agents: Expert, Motivator, and Mentor. The expert agent was designed to exhibit mastery in the subject area; the motivator agent was intended to represent a peer modeling self-efficacy support, and the mentor agent was designed to represent a guide or coach with advanced experience and knowledge to work collaboratively with the learners to achieve goals. They found that the pedagogical agents can enact human-like roles within a learning system in which learners are active meaning-makers of their environments, bringing their own perspectives and expectations to the learning context. Johnson and Lester (2016) also contend that with the advancement of research in A.I., pedagogical agents will have increasingly greater abilities to recognize when learners exhibit frustration, boredom, confusion, and states of flow, among others, and become emotionally more intelligent to comfort, support, and motivate learners (Johnson & Lester, 2016). In sum, pedagogical agents can personalize learning and assist learners and the teachers by taking on a range of roles from teaching and guiding, through collaborating and co-learning, to existing in the background to deepen student learning. Future pedagogical agents could provide highly customized support for all subject matters and expand its support to metacognitive and self-regulatory skills and beyond.

1.3.9.2 Personalized Learning Pedagogy

There is still debate on how personalized learning (P.L.) can be defined in the age of rapid advances in technology platforms and digital content (NFES, 2019). However, the most common definition refers to P.L. as learner-centered learning and learning that is differentiated and responsive to learners' readiness levels, interests, and learning profiles. The U.S. Department of Education, Office of Educational Technology (2017) provides the most cited definition. It refers to personalized learning as instruction in which the pace of learning and the instructional approach are optimized for each learner's needs. Furthermore, according to this definition, learning activities are meaningful and relevant to learners, driven by their interest, and often self-initiated. Consistent with these definitions, educators have long believed in a learning environment that meets each student at their level, challenges them with high expectations, increases student agency over their learning, and addresses the individual learner's needs. However, in a classroom with a large group of students, it has been difficult for educators to fully personalize learning and address students' individual needs. The advances in technology, namely big data, data analytics, availability of digital curricular content, learning spaces that combine formal and informal learning anyplace and at any time and with anything (e.g., IoT; A.R.) have made it possible for educators to personalize learning to assist students in constructing foundational knowledge at their level while teachers spend more time facilitating, coaching, and guiding them in their learning processes. In other words, the technology-supported personalized learning does not diminish the central role that teachers play, particularly in enhancing K-12 learners' ability to construct knowledge, develop cognitive skills (e.g., self-regulation, executive control, self-direction), and grow socially and emotionally. Teachers play critical roles in smart learning environments by actively using technology to support individual students' needs and interests. Figure 1.4 provides a conceptualization of personalized learning.

With the above-mentioned stipulation and the rise of big data, it is possible to record and interpret students' characteristics and real-time state in smart learning environments. The following four implementation goals can be considered for smart learning environments that utilize the support of smart and intelligent technology to pursue personalized learning (Boninger, Molnar, & Saldaña, 2019). Personalized learning should:

- *Provide regularly updated records of students' strengths, needs, motivations, and goals.* Detailed knowledge of students' needs, strengths, interests, and achievement of different competencies are needed to optimize learning experiences for each student. Smart learning environments create learner profiles for students. Using advanced A.I. technologies, the recorded data can be tracked, analyzed,

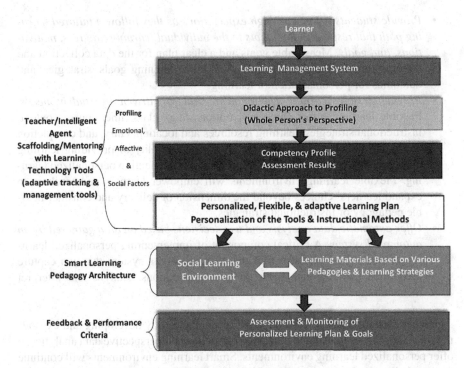

FIGURE 1.4   A conceptualization of personalized learning.

and updated to allow students to pursue personal interest areas and help teachers develop more challenging learning experiences for individual students.

- *Continuously assess students' progress toward their goals and give them credit when they achieve learning outcomes.* Meaningful and frequent formative data and automated analysis of multiple data sources are needed to provide immediate, targeted feedback to students and help teachers use formative assessments to modify and adjust instruction to learners' needs. The challenge with implementing this option is over-emphasizing quantitative methods of collecting assessment data (e.g., objective tests, grades) for predictive analytics. In other words, whereas big data methods can easily score objective tests, attitude, or interest surveys, facial expressions, even attach a "sentiment score" to individual student narrative response, it is less clear whether existing methods can identify issues such as the presence of or lack of empathy, interest, emotions in narrative data or assessing learning in open-ended performance tasks. Design, development, and implementation of new models, frameworks, and algorithms are needed to predict and measure performance-based learning, particularly in learning environments that new pedagogical learning models are used (Sin & Muthu, 2015). The application of learning analytics in assessments enabling automated, real-time feedback in social learning where students build knowledge collaboratively in their cultural and social settings through discourse or essay answers will impact how learning can be personalized.

- *Provide students with clear, high expectations as they follow a tailored learning path that responds and adapts to the individual learning progress, motivations, and goals.* Measurable goals and a clear plan for the data collection and defining timelines and expectations will allow aligning goals, strategies, and outcomes for personalization of learning.
- *Provide flexible learning environments that are driven by student needs.* Flexible learning offers choices or options (e.g., class time, course content, instructional strategies, learning resources and location, entry, and completion dates) in the educational environment. The possibility of providing learning choices to learners is a crucial component of implementing personalized learning. Flexible learning environments will empower learners and teachers to expand learners' choices beyond the dimension of delivery and consider flexible pedagogy or instructional strategies.
- *Offer automatic data analysis and use decisions based on data gathered by an automated system.* A critical component of implementing personalized learning is creating an infrastructure that ensures a data system that can capture predefined learning data from multiple applications, analyze and interpret them automatically.

Higher education and K-12 school districts are currently developing and expanding their approaches to using big data and predictive and perspective data analytics to offer personalized learning environments. Smart learning environments will continue facing challenges in developing a vision for the personalized learning programs, identifying types and specific uses of data, collecting meaningful macro (programs, schools, and districts level) and micro-level (course, subject area) learning data from multiple data systems while ensuring privacy, equity, social justice, and security, and grappling with how to handle structured versus unstructured data. Nevertheless, the advances in A.I., machine learning, big data, and data analytics will allow new and innovative perspectives on designing and implementing personalized learning.

1.3.9.3 Adaptive Learning

Adaptive learning refers to learning that uses intelligent or artificial technology to dynamically monitor student progress based on generating performance and engagement data and then modify instruction (Adams Becker et al., 2017) to offer proper adjustment strategies (Waters, 2014). In other words, adaptive learning is a technology-based educational method that uses computers as interactive teaching and training devices to provide individual learning programs/materials to learners based on the data gathered throughout the learning/training process. Personalized learning happens through adaptive technology. Adaptive learning technology focuses on the learner's goals and makes intelligent adjustments based on the learner's needs and needs via data analytics. In sum, according to the U.S. Department of Education, Office of Educational Technology (2017),

adaptive learning systems use a data-driven, and in some cases, a nonlinear approach to instruction and remediation. They dynamically adjust to student interactions and performance levels, delivering the types of content in an appropriate sequence that individual learners need at specific points in time to make progress.

Based on the definitions of personalized and adaptive learning, while the first embraces learners' needs and characteristics to offer learning choices, the latter uses learner interaction with the learning system (e.g., choosing learning goals, desired topics, time frame) to trigger an action and once the actions are triggered the systems tries to adapt and assign tasks to learners based on their progress (Vesin, Mangaroska, & Giannakos, 2018). Peng, Ma, and Spector (2019) proposed combining the core elements of personalized learning and adaptive learning. They offer the following features as core elements of both personalized and adaptive learning: "individual characteristics, individual performance, personal development, and adaptive adjustment" (p. 6). Peng et al. (2019) further explain that the first three elements correspond to three customized levels of personalized learning; the fourth refers to the adaptive adjustments strategy of teaching to achieve the three personalized levels. Thus, in a combined personalized adaptive learning pedagogy, the teaching strategies are adjusted based on the differences in individual learner characteristics, differences and changes in current individual learner performance, and differences and changes in learner personal development (Peng et al., 2019).

In sum, the main idea behind adaptive learning is to build specific context using user-centered analytics and create learning strategies that could guide learners to accomplish their goals.

1.3.9.4 Flexible or Open Learning Pedagogy

Flexible and open learning refers to an environment that is increasingly free from the limitations of time, place, and pace of learning. The flexibility further includes choices regarding the entry and exit points, selection of learning activities, communication methods, assessment tasks, and educational resources. Flexible, open learning pedagogy is a learner-centered strategy that aims to make learners more self-determined and independent, while teachers became more of learning facilitators. With the technological advances, flexible and open learning empowers learners and instructors to extend learning opportunities beyond the dimension of delivery to cover flexible pedagogy. Furthermore, anytime, anywhere, learning bridges formal and informal learning experiences possible through the effective use of technology. Therefore, digital learning modalities combined with performance-based progressions make it possible to meet each student's unique needs (iNACOL, 2016). Naidu (2017) offers a more detailed view of the flexible pedagogy by suggesting the following embedded key dimensions of learning and teaching:

1. *Learning experience design* where the design of learning experiences allows each learner to make the most of the learning opportunities.
2. *Learner-content engagement* where the learners' interaction and engagement with the subject matter are aligned with their interests, styles, time, place, and pace.
3. *Learner-teacher engagement* is about choices the learner has regarding the mode and methods of interacting with teachers and tutors.

4. *Learner-learner engagement* is the learners' choices regarding the mode and methods of interacting and engaging with their peers in small and large groups in online or offline educational settings.
5. *Learner engagement with the learning environment* where the learner has access to various ways of interacting with the learning environment (e.g., mobile, A.R., IoT).
6. *Learner engagement with assessment activities* where the learners may choose how to fulfill their assessment requirements.
7. *Learner engagement with feedback* where the learners have access to feedback on their learning and assessment requirements.
8. *Learner engagement with the institution* where the learners have choices regarding their engagement with the services of the education institution. (Naidu, 2017. p. 270)

Related to the concept of flexible learning is the movement in *Open Educational Resources* (O.E.R.). O.E.R. are teaching, learning, and research materials in any medium that reside in the public domain and are released under an open license that permits no-cost access, use, adaptation, and redistribution by others (Educause, 2018). O.E.R. covers a range of learning resources, including textbooks, curricula, video, audio, simulations, assessments, and many other contents used in education. O.E.R. provide up to date learning materials that can be tailored and adapted, revised, expanded, and translated, and shared with educators and learners. Therefore, O.E.R. support the practice of open and flexible learning by providing access to learning content for everyone (Educause, 2018).

In sum, flexible learning environments support student-centered pedagogical approaches. They encourage greater learner autonomy and empower learners to control their own learning.

1.3.9.5   Interest-Driven Learning

Interest-driven learning is a learning approach that views the learner as the main actor in the learning space, responsible for maintaining social relationships and creating meanings throughout the physical and virtual context. It uses interest [both individual (personal) interest and situational interest] as a motivator for learning. Personal interest is an attraction that emerges from the learner's experiences. Situational interest arises from the situation rather than the learner experience. Interest-driven learning (elicited by the content of the learning task) engages learners in certain activities to seek specific knowledge or skills. The benefits of interest-driven learning as a motivator for learning activities are associated with the learner's perception that certain knowledge or skills are useful to pursue an interest (a mastery goal). It is also connected with expending more effort and persists longer at the tasks. Self-determination theory demonstrates that intrinsic motivation leading to persistence cannot emerge unless a person has a sense of autonomy (Ryan & Deci, 2006).

The context can also drive learner interest. Examples of context motivators are social context (socially and culturally relevant context), extrinsic rewards, level of structure and choice, and task difficulty level. Thus, the ideal smart learning environment considers learners' view of whether a learning activity/task or problem is relevant, achievable (competence for completing the task/activity), can be done collaboratively and by participating in a social group (conversation with others) and engaging in reflection, self-assessment, and self-organization (Spector, 2014; Gros, 2016). Therefore, a learning environment designed based on interest-driven pedagogy recognizes that when the learners see the authentic use of knowledge or skills beyond just the structure of the educational settings, they will be motivated to engage in learning activities.

1.3.9.6 Collaborative Learning Pedagogy

Collaborative learning is learning that occurs in a situation that two or more people learn or attempt to learn something together. It is grounded in social constructivist learning theory, which emphasizes that learning and knowledge construction is affected by interaction and collaboration (Vuopala, Hyvönen, & Järvelä, 2016; Palincsar, 1998). Collaborative learning includes learners' mutual engagement in a joint effort to construct knowledge and solve problems together. Given the pedagogies of personalized, adaptive, and flexible learning, it may appear that in smart learning environments, the focus of learning is more on the individual learners who navigate through their own learning path. However, people do not live and learn in isolation from others (Spector, 2014). Moreover, the positive impact of social interactions on individual learning is highlighted in the literature.

Therefore, in technology-supported smart learning environments, an active and engaged learner continuously interacts with others formally during the completion of learning activities and tasks and informally, particularly when communicating with experts and peers using social media. In other words, electronic devices and social media create opportunities for learners to learn collaboratively and allow them to share ideas, thoughts, and resource materials. Furthermore, virtual learning communities, especially in higher education, are formed in cyberspaces, making it easier to collaborate with geographically dispersed learners. Moreover, research has shown that collaboration and group work can support deep learning pedagogies (Baeten, Kyndt, Struyven, & Dochy, 2010), enhance learners' engagement in various forms of inquiry-oriented activities (Kuhn, Black, Keselman, & Kaplan, 2000), and engage students in investigations of personally relevant questions in collaboration with their peers.

1.3.9.7 Engaged and Meaningful Learning

Engaged and meaningful learning is learning that engages the learner in solving real-world problems/scenarios and with computer-supported and I.T. that provides the necessary learning guidance, hints, supportive tools, or learning suggestions in the right place, at the right time, and in the right form (Gros, 2016). An engaging learning environment emphasizes motivating and sustaining learner interest by offering relevant and real-world tasks and creating a conversational, reflective, and innovative learning space (Spector, 2014). Without learning and practicing in meaningful, authentic environments (interest-driven), engaging in a dialogue with peers (collaborative learning), having an opportunity to self-assess and reflect, and developing innovative ideas (deep learning), students will not be able to apply the knowledge to solve problems. Thus, engaged learning further emphasizes learner responsibility, metacognitive skills, and dialogical and collaborative teaching and learning models.

1.3.9.8 Formal and Informal Learning (Anywhere, Anytime, Anything and Any Place Learning)

Learning occurs not just in classrooms but also in the home, workplace, playground, library, museum, nature center, and learners' daily interactions with others and during leisure activities (Liu, Huang, & Wosinski, 2017). Informal learning is often initiated by the students and motivated by their own interest. Smart learning environments are not restricted to formal settings; they connect formal with informal learning and bridge both types of learning by interconnecting different technologies (e.g., IoT, A.R., wearable, mobile devices) that enable linking formal and informal learning across spaces. For example, a personalized smart learning environment gathers data about the students' actions, interactions, and interests during the learning process, tracks the student learning by the help of data analytics or measuring tools, models the students' participation in different activities and resources and their progression, and then intervenes and interacts with the student by offering and providing appropriate resources and activities. Thus, integration and convergence of diverse embedded sensors and wireless devices and various interconnected technologies support formal and informal learning seamlessly. Informal learning, therefore, can be planned learning by the teacher/designer, an adaptive learning system, or self-initiated and driven by the student's interest and motivation.

1.3.9.9 Generative Learning

Generative learning involves actively constructing meaning from to-be-learned information by mentally reorganizing it and integrating it with one's existing knowledge (Fiorella & Mayer, 2016). This form of active, cognitive learning involves creating and curating and allowing learners to engage in more complex learning activities. Such a learning approach supports conversation with others, reflection, innovation, and self-organization (Spector, 2014). Thus, generative learning depends not only on utilizing instructional methods and how the information is presented to learners but also on how learners try to make sense of it. For example, in project-based learning, students learn by identifying, analyzing, synthesizing, and generating possible solutions to represent their understanding of the problem.

Adopting proper instructional strategies in the smart learning environment is critical to enhancing generative learning. Fiorella and Mayer (2016) recommend a series of instructional strategies that strongly support generative learning pedagogies. The first four strategies recommend summarizing, mapping, drawing, and imagining the information. These strategies allow the learner to select the most relevant information in the task or the problem and organize them by building connections among the elements of information selected. The second four strategies are self-testing, self-explaining, teaching, and enacting. These strategies help the learner elaborate upon the materials, generate answers to the questions, explain it to others, and act out the materials with concrete objects (Fiorella & Mayer, 2016). However, as emphasized by Fiorella and Mayer, although instructional strategies are important in enhancing generative learning, what the learner can do (metacognitive strategies) to foster generative learning is also important.

Nonetheless, generative learning pedagogy requires assessing meaningful or generative learning outcomes, such as problem-solving, critical thinking, cognitive, and metacognitive thinking skills. Thus, it is important to develop measures of processes and strategies used during learning in the smart learning environment. The process measures can be used to analyze the quality of what students generate during learning (Fiorella & Mayer, 2016). Additionally, according to generative learning theory, the most important difference among students is domain-specific prior knowledge and general ability, metacognition skills, or motivation. Thus, smart learning environments should not only create an environment that promotes generative learning but also properly assesses and tracks meaningful learning.

1.3.9.10 Assessment that Uses Continuous Learning

Assessment that uses continuous learning refers to advanced learning systems that can review, monitor, and track learners' learning progress and provide guidance (i.e., feedback, hints) towards achieving their learning goals (Hwang, 2014; Liu et al., 2017). In other words, smart learning environments offer immediate and adaptive support to learners by instant analyses of learner's learning performance, learning behaviors, profiles, personal factors to provide learning guidance, feedback, and learning tools and materials (formative assessment), based on the learners' needs. The formative and summative feedback includes electronic assessments, adaptive testing, and feedback methods that are personalized, automated, real-time, evidence based, and data-driven.

However, as indicated earlier, while in smart learning environments, assessment of learning outcomes are designed as adaptive (using computer algorithms to offer computer adaptive testing), delivered online, and automatically scored to provide immediate feedback, they are still based on the science of assessment that is more grounded on discrete knowledge-based items (item formats such as multiple-choice, selected response), most of which are somewhat divorced from the performance-based learning environments that make up the contexts students encounter (Shute, Leightonb, Jangc, & Chu, 2016). The design, development, and implementation of new models, frameworks, and algorithms are needed to predict and measure performance-based learning, particularly in learning environments that new pedagogical learning models are used (Sin & Muthu, 2015). Thus, along with the smart pedagogies, assessment strategies and curricula need to change, permitting greater ecological validity and feedback to students related to breadth and depth of knowledge and skills (e.g., critical thinking, creativity, collaboration, problem-solving) as well as the assessment of multidimensional learner characteristics (cognitive, metacognitive, and affective) using authentic, digital task (Shute et al., 2016).

Ironically, assessment of learning often lags behind the technology used to support it (Kinshuk et al., 2016; Spector, 2014). While smart learning environments are used to accomplish a range of learning outcomes, generative learning outcomes such as creating simulations, running a live simulation, analyzing problems or case studies, developing solutions or products in a collaborative learning environment require an assessment system that can analyze complex learning outcomes both formatively and summatively. The smart pedagogies should focus on assessment strategies that measure the content of the learning domain and how knowledge of content is applied, manipulated, or processed.

1.3.10 Smart Education and Collaboration with the Community

The smart education system does not work independently. It operates in a *collaborative environment*. In a smart education system, school districts or universities collaborate with community groups, cultural institutions, businesses, and city agencies to provide open, flexible, personalized, and adaptive education.

Creating an environment of collaboration is one of the main differences between smart and traditional education. The community collaboration in smart education considers students, families, and communities as resources and collaborators in improving educational systems and enacting new roles, power dynamics, and interactions among various members of the community. Establishing interaction and collaboration between different stakeholders optimizes education and emphasizes the importance of using multiple elements of smart learning and smart pedagogies. The concept of involvement and collaboration with the community in a smart education system mirrors the concept of a smart city. Nikolov et al. (2016) define a smart city as an "effective integration of physical, digital, and human systems in building an environment to deliver a sustainable, prosperous and inclusive future for its citizen" (Nikolov et al., 2016. p. 339). Similar to the concept of smart city, the smart education system is a combination of advanced and immersive technology infrastructure, smart learning environments, and smart pedagogies that enable the education system and various community organizations to work together by sharing responsibilities and authority for decisions on operations, policies, or actions of the education system. Clearly, advanced and ubiquitous technologies play an important role in connecting and facilitating internal and external collaboration to achieve shared educational goals. A collaborative smart education system can apply collective intelligence for innovative solutions to community problems and can offer shared responsibilities that ultimately build trust and confidence in education systems. As with smart cities, information technologies, various smart devices, cloud computing, the IoT, big data, and geographical information facilitate community collaboration in the smart education system.

Examples of community collaboration in the smart education system are abundant. Community connects and creates opportunities for experiential, authentic learning. It creates opportunities for students to co-construct knowledge with their communities. Local businesses, subject matter experts, and mentors link formal curriculum with the outside world and assist in designing engaging learning experiences outside the classroom, and technology can facilitate this connection. Project-based and problem-based learning pedagogies (deep learning) gain their powers when connected to the community and help learners work on real-world projects or solve community problems. When students, teachers, and communities work together, they can use advanced and emerging technologies to design and facilitate change locally. More broadly, community collaboration and engagement in a smart education system makes it possible for the community at large to play an important role in taking greater responsibilities in reducing socioeconomic inequality, supporting students' social and emotional needs, assisting in lowering the overloaded responsibilities of teachers, increasing technology access, and actively participating in protecting the safety, welfare, health, and well-being of students.

In sum, community collaboration is not an option but a critical element of smart education. It mobilizes resources and support and demands for an effective and accountable school system. It creates opportunities for students to co-construct knowledge with their communities. The development of a smart education ecosystem enables community presence at all levels (local, regional, national) and in spaces of coexistence with a broad and integrated approach. The new technology serves to increase the effectiveness of the vast array of operations that occur in it.

1.3.11 IMPLEMENTATION SUPPORT AND FACILITATION IN SMART EDUCATION SYSTEM

A smart education system cannot operate without robust *support and facilitation*. Technology, instructional design, universal design, and teacher's support are needed to provide for students' success in the smart education system.

The successful implementation of smart education depends highly on strategies used to support the operation of advanced and immersive computing infrastructure. First, the education institutions should invest in advanced and immersive computing infrastructure (both the first and the second layer of computing infrastructure discussed earlier). In other words, the investment should first focus on systems and services (first layer) that support the delivery of advanced computing (second layer). Secondly, particular attention should be paid to providing proper support for the evolution of big data and data analytics, hardware, software, facilities, and services optimized for data-intensive computing. Visionary leadership is needed to collaborate with the larger community and create roadmaps that allow making strategic and collaborative decisions to support and facilitate the transformation of the old computing system to the new, advanced, and futuristic one. Thirdly, the advanced computing infrastructure should support and require applications that broaden the equity, accessibility, and utility of large-scale platforms and cost-effective use of cloud computing and execution of data analytics with specific attention on security and privacy.

The implementation of first and second layers of computing infrastructure and continuously evaluating future advanced computing capabilities are challenging and require institutions to look into the future and meet the demand for computational resources that move away from a traditional education system to a smart and intelligent education system.

The implementation of a smart learning environment and smart pedagogies requires a different set of strategies and support. As indicated earlier, the advanced technologies and smart learning environments and pedagogies are interdependent (Cros, 2016b). Technology creates the context in which learning takes place. The features of smart learning environments (e.g., adaptive, flexible, and personalized learning, a mixture of formal and informal) and related pedagogies (e.g., deep, learner-directed, interest-driven, collaborative, generative, meaningful learning) call for support and facilitation that prepare teachers and teacher educators for the following set of knowledge and skills.

1.3.12 NEW ROLES AND RESPONSIBILITIES FOR TEACHERS

Teachers should be prepared for a learning environment where learning happens anyplace, anytime, with anything, and students can participate simultaneously in the same activity in a different space, time, and social settings. The new learning process requires learning materials that are designed, developed, tested, and available and accessible from any devices. Furthermore, teachers must be prepared to take the role of a guide, facilitator, and mediator of learning that occurs formally (in physical and digital worlds) and informally (outside of the classroom) across a combination of locations, times, technologies, and social settings. In their new roles, teachers should realize that technology is the context of learning and is the necessary condition to predict and advise learners' needs and to provide a learning system that is situated in the real-world contexts; the tasks are adapted to individual learners; guidance and feedback is focused on the development of learners' capability to learn, create, and assume agency in their learning. The concept of teachers as co-designers is particularly critical in smart learning environments as these environments require a significant reconceptualization of teaching practices, interactions between teachers and students, and the role of the physical and virtual spaces.

1.3.13 TEACHERS AS CO-DESIGNERS OF LEARNING MATERIALS AND RESOURCES

The success of smart learning depends on the design of its learning processes, learning materials, and resources. The field of learning design offers a set of principles, tools, systems, and models that can empower educators to design challenging, complex, and real-world learning tasks, or problems. Learning designers further pay close attention to learning outcomes, learning activities, learning contexts, learning resources, learning tools, learning community, scaffolds, assessment, and evaluation as key elements of learning environments. Additionally, the learning design principles and procedures offer an iterative and collaborative process in which the teacher, as a co-designer of learning, incorporates learners' knowledge and skills, backgrounds, interests, preferences as key elements in creating and supporting learning through continuous tracking of learner's activities and offering immediate feedback. Thus, educators must be supported to become learning designers who are well aware of learning theories and new pedagogical models related to technology-enhanced learning.

1.3.14 UNIVERSAL DESIGN FOR LEARNING AND ACCESSIBILITY

Universal design for learning (U.D.L.) promises to build an architecture when developing learning materials/contents (Meyer, Rose, & Gordon, 2014). Consistent with the concepts of personalized, flexible learning, U.D.L. suggests a learning design that focuses on multiple representations (using a variety of methods to represent the knowledge) and customization of knowledge, various means of actions and expression (allowing students to choose how they demonstrate their knowledge and develop abilities to self-monitor), and multiple means of engagement (multiple ways to motivate learners to engage in learning) (Meyer, Rose, & Gordon, 2014). Thus, U.D.L. supports the flexible design of learning and offers a wide array of opportunities that improve accessibility. Figure 1.5 provides a conceptualization of smart education with all its interrelated elements explained in the previous sections.

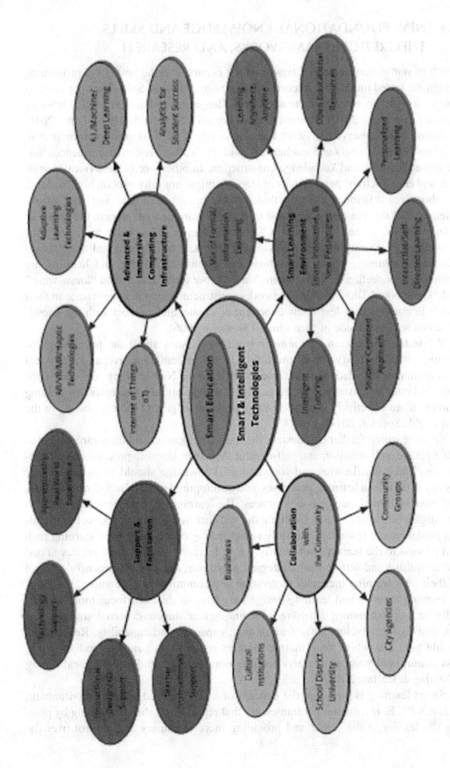

FIGURE 1.5 Smart education: A conceptual framework.

1.4 NEW FOUNDATIONAL KNOWLEDGE AND SKILLS, THEORETICAL FRAMEWORKS, AND RESEARCH

Much of our understanding of how learning occurs, and the best ways of learning design are based on learning theories and learning sciences. Some learning theories and learning sciences continue to address challenges related to learning and learning design (Anderson, 2010). However, as Anderson (2010) and others (Gros, 2016; Siemens, 2006) observed, network learning in which I.C.T. is used to promote connections among learners and teachers and with the learning resources influences how we define learning and knowledge construction. In other words, behavioral, cognitive, and constructivist perspectives of learning might not fully explain highly social, collaborative, relational learning that occurs through dialogue and interactions (Siemens, 2006). The new conception of network learning that refers to the theory of "*connectivism*" coined by Siemens (2006) characterizes knowledge as connections between a network of humans (individuals, groups, communities) and non-humans (resources, systems, ideas). According to Siemens, knowledge and learning are defined by connections; "know where," and "know who," rather than "know what," and "know how." Thus, learning networks are structures that learners create in their minds to continuously learn new experiences, create, and connect with the knowledge that resides outside of their minds (Siemens, 2006).

While the conception of *connectivism* raises issues about the psychology of learning, it lacks clarity about how dialogical, social, collaborative, and reciprocal construction of knowledge plays a role in our learning. New learning concepts might provide better opportunities for researchers to develop new strategies for helping learners more effectively and efficiently gain knowledge and solve problems in the real world (Spector, 2014).

Smart learning further emphasizes the individual learner and their capabilities to self-manage, self-monitor, and self-control their own learning processes and experiences. Thus, the theories and foundational knowledge should focus on *learner agency* in their own learning processes, and developing *learners' self-determination, self-control* through personal experiences. The learning design theories should also investigate how teachers can facilitate the self-directed learning process by providing guidance and resources and fully relinquishing ownership of the learning path and process to the learner, particularly young learners who have not yet developed metacognitive and self-control strategies. Moreover, even though learners' control of their own learning increases motivation and commitment to learning, the focus of learning theories and learning design may have to shift and focus more specifically on *self-determined learning or heutagogy*, a student-centered strategy that emphasizes the development of autonomy, capacity, and capability. Researchers should note that while more mature learners require less guidance and structure, less mature and younger learners require more teacher guidance and scaffolding (Canning & Callan, 2010).

Smart learning is based on the concept of a personalized learning environment (P.L.E.). P.L.E. is a theoretical framework that represents a shift in learning by placing the learner at the center and providing more autonomy and control over the learning experience. A well-defined theoretical framework is needed to take advantage of affordances of technology to specifically design an external network that can incorporate new knowledge nodes, identify connections between different knowledge nodes, and locate the knowledge note that can help achieve better results (Gros, 2016). This theoretical foundation should define learning design principles that support multidirectional and multimodal learning in which learning takes place in different socio-cultural contexts formally and informally, considers learner's personal and cultural experiences, and reduces the barriers to participation in various learning opportunities. It should provide a foundation for educators to develop transformative practices and understand more about the context in which this new pedagogy is formed.

Guidance and scaffolding are critical components of smart learning environments. To reduce the teacher's cognitive load and benefit from intelligent technology, there is a need to expand the foundational knowledge about *pedagogical agents*. An animated virtual agent can play the role of a tutor, consultant, companion, or confidant while the learner is engaged in learning. Research on pedagogical agents needs to understand how and when, and in what forms, pedagogical agents should be available to learners and whether the learner requires to have developed metacognition and self-awareness before benefiting from the virtual agent.

The increasing focus on digitization and adoption of digital technologies offers personalized, adaptive, and immersive learning results in collecting a vast amount of data to create learner profiles in a new and unique way. Such user profiles that provide insights into their preferences, beliefs, and routines raise *privacy and ethical issues*. More advanced knowledge and best practices are needed to protect personal data against unauthorized processing, create strategies for developing unbiased categorization of data for data analytics, and to implement appropriate technical and organizational measures, including deleting personal data when it is no longer necessary and many other ethical and privacy issues.

Finally, as discussed earlier, a smart learning environment must offer instant and adaptive support to learners by immediate analyses of the needs of individual learners from different perspectives (e.g., learning performance, learning behaviors, profiles, personal factors). Furthermore, the concept of personalized learning support requires guidance and feedback based on the learners' needs. To make these happen, a *smart assessment system* is needed. However, despite the advances in A.I. learning, design theories and principles regarding assessment practices in educational settings have not changed. Smart learning needs to develop foundational knowledge regarding detecting meaningful patterns from the learning processes for understanding the student learning activities and improving learning outcomes.

To move beyond traditional forms of assessment, new methods and algorithms are needed to combine *assessment of learning outcomes/products and process of learning*. Furthermore, a balance evidence-based, real-time assessment (especially self-assessment) with intelligent digital systems designed to foster critical thinking and problem-solving should be developed. Strategies for managing and tracking learning process data should be extended to inform the design of smart learning environments.

1.5 CONCLUSION

The chapter analyzed and discussed the concepts of smart education, smart learning, smart learning environment, and related pedagogies. It further examined and explored the various elements of advanced and immersive technology infrastructure, features of smart learning environments, and collaborative approaches to supporting and implementing smart education systems. The chapter attempted to explain how smart technologies and shifts in learning and instructional strategies disrupt traditional education and move it to a smarter learning environment. It concluded that technological development and the establishment of smart learning have resulted in the needs for new foundational knowledge and skills, theoretical frameworks, and field of research.

REFERENCES

Adams Becker, S., Cummins, M., Davis, A., Freeman, A., Hall Giesinger, C., & Ananthanarayanan, V. (2017). *N.M.C. Horizon Report: Higher Education Edition*, The New Media Consortium.

Adi, E., Baig, Z., & Hingston, P. (2017). Stealthy Denial of Service (DoS) attack modelling and detection for HTTP/2 services, *Journal of Network and Computer Applications, 91*, 1–13. https://doi.org/10.1016/j.jnca.2017.04.015

Adi, E., Anwar, A., Baig, Z., & Zeadally, S. (2020). Machine learning and data analytics for the IoT. *Neural Computing and Applications, 32*, 16205–16233.

Alammary, A., Alhazmi, S., Almasri, M., & Gillani, S. (2019). Blockchain-based applications in education: A systematic review. *Applied Science, 9*(12), 2400.

Aldowah, H., Rehman, S. U., Ghazal, S., & Umar, I. N. (2017). Internet of Things in higher education: A study on future learning. *Journal of Physics: Conf. Series, 892*.

Anderson, T. (2010). Theories for learning with emerging technologies. In, G. Veletsianos (Ed.). *Emerging technologies in distance education* (pp. 23–40). Athabasca University Press.

Andujar, J. M., Mejias, A., & Marquez, M. A. (2011). Augmented reality for the improvement of remote laboratories: an augmented remote laboratory. *IEEE Transaction on Education 54*(3), 492–500.

Annenberg Institute for School Reform (Voices in Urban Education) (2010). *Building smart education system*. Rhode Island: Annenberg Institute for School Reform at Brown University Pub.

Baeten, M., Kyndt, E., Struyven. K., & Dochy, F. (2010) Using student-centred learning environments to stimulate deep approaches to learning: Factors encouraging or discouraging their effectiveness. *Educational Research Review, 5*(3): 243–260.

Boninger, F., Molnar, A., & Saldaña, C. M. (2019). *Personalized learning and the digital privatization of curriculum and teaching*. National Education Policy Center. Retrieved from http://nepc.colorado.edu/publication/personalized-learning.

Botella, C., Garcia-Palacios, A., Baños, R. M., & Quero, S. (2009). Cybertherapy: advantages, limitations, and ethical issues. *PsychNology Journal, 7*, 77–100.

Brown, M., McCormack, M., Reeves, J., Brook, D. C., Grajek, S., Alexander, B., Bali, M., Bulger, S., Dark, S., Engelbert, N., Gannon, K., Gauthier, A., Gibson, D., Gibson, R., Lundin, B., Veletsianos, G., & Weber, N. (2020). *2020 Educause Horizon Report Teaching and Learning Edition*. EDUCAUSE. Retrieved December 27, 2020 from https://www.learntechlib.org/p/215670/.

Canning, N., & Callan, S. (2010). Heutagogy: Spirals of reflection to empower learners in higher education. *Reflective Practice, 11*(1), 71–82.

Christou, C. (2010). Virtual Reality in Education. In A. Tzanavari, & N. Tsapatsoulis (Eds.), *Affective, interactive and cognitive methods for e-learning design: Creating an optimal education experience* (1st ed., pp. 228–243). I.G.I. Global.

Coelho, O. B., & Silveira, I. (2017). *Deep learning applied to learning analytics and educational data mining: A systematic literature review.* In *Brazilian Symposium on computers in education (simposio Brasileiro de Informatica na educaçao-SBIE)*, 28, p. 143.

Darling-Hammond, L., Oakes, J. (2019). *Preparing teachers for more in-depth learning.* Harvard Education Press.

Educause (2018). 7 Things you should know about …, *Educause Learning Initiative (E.L.I.)*, EDUCAUSE Learning Initiative | EDUCAUSE.

Fiorella, L., & Mayer, R. E. (2016). Eight ways to promote generative learning, *Educational Psychology*, *28*, 717–741.

Freeman, J., Lessiter, J., Pugh, K., & Keogh, E. (2005). *When presence and emotion are related, and when they are not*, in *Proceedings of the Conference at Presence 2005*, London. Retrieved from http://www.temple.edu/ispr/prev_conferences/proceedings/2005/freeman,%20lessiter,%20pugh,%20keogh.pdf.

Furht, B. (2011). *Handbook of augmented reality.* Springer.

Galletta, D., & Zhang, P. (2014). Human-computer interaction and management information systems: Applications. In *Advances in Management Information Systems*, Volume 6. Routledge.

Greer, D., Crutchfield, S., & Woods, K. (2013). Cognitive Theory of multimedia learning, instructional design principles, and students with learning disabilities in computer-based and online learning environments. *The Journal of Education*, *193*(2), 41–50. Retrieved November 25, 2020, from http://www.jstor.org/stable/24636945

Gros, B. (2016a). The design of smart educational environments. *Smart Learning Environments*, *3*(15), 1–11.

Gros, B. (2016b). The dialogue between emerging pedagogies and emerging technologies. In B. Gros, & M. M. Kinshuk (Eds.), *The future of ubiquitous learning* (pp. 3–23). Springer.

Gwak, D. (2010). The meaning and predict of Smart Learning, Smart Learning Korea *Proceeding, Korean e-Learning Industry Association.*

Hoel, T., Mason, J. (2018). Standards for smart education – towards a development framework. *Smart Learning Environments*, *5*(3), 1–25.

Hwang, G. J. (2014). Definition, framework, and research issues of smart learning environments-a context-aware ubiquitous learning perspective. *Smart Learning Environments*, *1*(1), 1–14.

Ijsselsteijn, W., De Ridder, H., Freeman, J., Avons, S. E., & Bouwhuis, D. (2001). Effects of stereoscopic presentation, image motion, and screen size on subjective and objective measures of presence. *Presence*, *10*, 298–311.

iNACOL (2016). What is personalized learning? Retrieved from http://www.inacol.org/news/what-is-personalized-learning/.

Jeong, H., & Hmelo-Silver, C. E. (2016). Seven affordances of computer-supported collaborative learning: how to support collaborative learning? How can technologies help? *Educational Psychologist*, *51*(2), 247–265.

Jeong, H., & Hmelo-Silver, C. E. (2012). Technology supports in CSCL. In J. van Aalst, K. Thompson, M. J. Jacobson, & P. Reinmann (Eds.), *The future of learning* (pp. 339–346), Proceedings of the 10th International Conference of the Learning Sciences (ICLS 2012)—Volume 1, Sydney. Society of the Learning Sciences.

Johnson, W. L., & Lester, J. C. (2016). Face-to-face interaction with pedagogical agents, twenty years later. *International Journal of Artificial Intelligence in Education*, *26*, 25–36.

Kamarainen, A. M., Metcalf, S., Grotzer, T., Browne, A., Mazzuca, D., Tutwiler, M. S., & Dede, C. (2013). EcoMOBILE: Integrating augmented reality and probeware with environmental education field trips. *Computers & Education*, *68*, 545–556.

Khaddage, F., Müller, W., & Flintoff, K. (2016). Advancing mobile learning in formal and informal settings via mobile app technology: Where to from here, and how? *Educational Technology & Society, 19* (3), 16–26.

Kim, T., Cho, J. Y., & Lee, B. G., (2013). Evolution to smart learning in public education: a case study of Korean public education. In L. Tobias, R. Mikko, L. Mart, and T. Arthur (Eds.), *Open and social technologies for networked learning*, Berlin Heidelberg: Springer, pp. 170–178.

Kim, Y., & Bayler, A. L. (2016). Research-based design of pedagogical agent roles: A review, progress, and recommendations. *International Journal of Artificial Intelligence in Education, 26*, 160–169.

Kinshuk, Chen, N.-S., Cheng, I.-L., & Chew, S. W. (2016). Evolution is not enough: Revolutionizing current learning environments to smart learning environments. *International Journal of Artificial Intelligence in Education, 26*, 561–581.

Kuhn, D., Black, J., Keselman, A., & Kaplan, D. (2000). The development of cognitive skills to support inquiry learning. *Cognition and Instruction, 18*(4), 495–523.

Lee, E. A.-L., Wong, K. W., & Fung, C. C. (2010). How desktop virtual reality enhance learning outcomes? A structural equation modeling approach. *Computers & Education, 55*, 1424–1442.

Liu, D., Huang, R., & Wosinski, M. (2017). Characteristics and framework of smart learning. In D. Liu, R. Huang, & M. Wosinski (Eds.), *Smart learning in smart cities* (pp. 31–48), Springer Nature.

Locker, L., & Patterson, J. (2008). Integrating social networking technologies in education: A case study of a formal learning environment. *2008 Eighth IEEE International Conference on Advanced Learning Technologies* (pp. 529–533). Santander.

Lozano, D. C. (2018). *Top emerging technology trends and their impact on L&D*. SweetRush Inc. eBook. Retrieved from https://sweetrush.com/ebook/Emerging_technology-trends.pdf

Lozano, J. C. (2020). *Hatoff to Adaptive Learning: Tailing corporate training for each learner*. SweetRush Inc. eBook. Retrieved from info@sweetrush.com.

Makransky, G. Terkildsen, T. S., & Mayer, R. E. (2019). Adding immersive virtual reality to a science lab simulation causes more presence by less learning. *Learning, and Instruction, 60*, 225–236.

Maseleno, A., Sabani, N., Huda, M., Ahmad, R., Jasmi, K. A., & Basiron, B. (2018). Demystifying Learning Analytics in Personalized Learning. *International Journal of Engineering & Technology, 7*(3), 1124–1129.

Mayer-Schönberger, V., & Cukier, K. (2013). *Big data: A revolution that will transform how we live, work, and think*. Houghton Mifflin Harcourt.

Meyer, A., Rose, D. H., & Gordon, D. (2014). *Universal design for learning: Theory and Practice*. CAST Professional Publishing.

Milgram, P., & Kishino, A. F. (1994). Taxonomy of mixed reality visual displays, *IEICE Transactions on Information and Systems, E77-D*(12), 1321–1329.

Miraoui, M. (2018). A context-aware smart classroom for enhanced learning environment. *International Journal of Smart Sensing and Intelligent Systems, 11*(1), 1–98.

Naidu, S. (2017). How flexible is flexible learning, who is to decide and what are its implications? *Distance Education, 38*(3), 269–272.

National Forum on Education Statistics (NFES). (2019). *Forum Guide to Personalized Learning Data (NFES2019160). U.S. Department of Education*. National Center for Education Statistics.

Nikolov, R., Shoikova, E., Krumova, M., Kovatcheva, E., Dimitrov, V., & Shikalanov, A. (2016). Learning in a smart city environment. *Journal of Communication & Computer, 13*, 338–350.

Nuffield Council on Bioethics. (2018). *Bioethics Briefing Notes: Artificial Intelligence (A.I.) in Healthcare and Research*. Retrieved from http://nuffieldbioethics.org/wp-content/uploads/Artificial-Intelligence-AIin-healthcare-and-research.pdf.

Orman, E. Price, H., & Russell, C. (2017) Feasibility of using an augmented immersive virtual reality learning environment to enhance music conducting skills. *Journal of Music Teacher Education*, *27*(1), 24–35 Retrieved from http://journals.sagepub.com/doi/pdf/10.1177/1057083717697962.

Palincsar, A. S. (1998). Social constructivist perspectives on teaching and learning. *Annual Review of Psychology*, *49*, 345–375.

Parsons, T. D., & Rizzo, A. A., Rogers, S., & York, P. (2009). Virtual reality in pediatric rehabilitation, *Developmental Neurorehabilitation*, *12*(4), 224–238.

Peng, H., Ma, S., & Spector, J. M. (2019). Personalized adaptive learning: an emerging pedagogical approach enabled by a smart learning environment. *Smart Learning Environment*, *6*, 9.

Petersson, D. (2020). *A.I. vs. machine learning vs. deep learning: Key differences*, Retrieved from TechTarget, https://searchenterpriseai.techtarget.com/tip/AI-vs-machine-learning-vs-deep-learning-Key-differences.

Picciano, A. G. (2012). The evolution of big data and learning analytics in American higher education. *Journal of Asynchronous Learning Network*, *16*(3), 9–20. 16.

Ryan, R. M., & Deci, E. L. (2006). Self-regulation and the problem of human autonomy: Does psychology need choice, self-determination, and will? *Journal of Personality*, *74*(6), 1557–1585.

Schubert, T., Friedmann, F., & Regenbrecht, H. (2001). The experience of presence: Factor analytic insights. *Presence*, *10*, 266–281.

Shute, V. J., Leightonb, J. P., Jangc, E. E., & Chu, M-W. (2016). Advances in science of assessment, *Educational Assessment*, *21*(1), 34–59.

Siemens, G. (2006). Knowing knowledge. Retrieved from http://www.elearnspace.org/KnowingKnowledge_LowRes.pdf.

Siemens, G., & Gasevic, D. (2012). Guest editorial-learning and knowledge analytics, *Educational Technology & Society*, *15*(3), 1–2.

Sin, K., & Muthu, L. (2015). Application of big data in education data mining and learning analytics – A literature review. *ICTACT Journal n Soft Computing*, *5*(4), 1035–1049.

Slater, M., & Wilbur, S. (1997). A framework for immersive virtual environments (FIVE): speculations on the role of presence in virtual environments. *Presence*, *6*, 603–616.

SoLAR (Society for Learning Analytics Research. (2011). *Open learning analytics: An integrated and modularized platform. Proposal to design, implement, and evaluate an open platform to integrate heterogeneous learning analytics techniques*, SoLAR.

Spector, M. J. (2014). Conceptualizing the emerging field of smart learning environments. *Smart, Learning Environments*, *1*(2). 1–10.

Stanford University. (2016). *One-hundred-year study on artificial intelligence* (AI100). Retrieved from https://creativecommons.org/licenses/by-nd/4.0/.

Tenzin, D., Lemay, D. J., Basnet, R. B., & Bazelais, P. (2020). Predictive analytics in education: A comparison of deep learning networks. *Education and Information Technologies*, *25*, 1951–1963.

Tran, B. X., Vu, G. T., Ha, G. H., Vuong, Q.-H., Ho, M.-T., Vuong, T.-T., La, V.-P., Ho, M.-T., Nghiem, K.-C. P., Nguyen, H. L. T., Latkin, C. A., Tam, W. W. S., Cheung, N.-M., Nguyen, H.-K. T., Ho, C. S. H., & Ho, R. C. M. (2019). Global evolution of research in artificial intelligence in health and medicine: A bibliometric study. *Journal of Clinical Medicine*, *8*(3), 360.

Turner, W. A., & Casey, L. (2014). Outcomes associated with virtual reality in psychological interventions: Where are we now? *Clinical Psychology Review*, *34*(8), 634–644.

U.S. Department of Education, Office of Educational Technology. (2017). *What is personalized learning*? Retrieved from https://medium.com/personalizing-the-learning-experience-insights/what-is-personalized-learning-bc874799b6f.

Uskov, V. L., Bakken, J. P., & Aluri, L. (2019). *Crowdsourcing-based learning: The effective smart pedagogy for STEM Education. IEEE Global Engineering Education Conference (EDUCON)*, Dubai, United Arab Emirates, pp. 1552–1558. doi:10.1109/EDUCON.2019.8725279.

Vesin, B., Mangaroska, K., & Giannakos, M. (2018). Learning in smart environments: user-centered design and analytics of an adaptive learning system. *Smart Learning Environment*, *5*, 24.

Vuopala, E., Hyvönen, P., & Järvelä, S. (2016). Interaction forms in successful collaborative learning in virtual learning environments. *Active Learning in Higher Education*, *17*(1), 25–38.

Waheed, H., Hassan S-U., Aljohani, N. R., Hardman, J., Alelyani, S., & Nawaz, R. (2020). Predicting academic performance of students from V.L.E. big data using deep learning models. *Computers in Human Behavior*, *104*, 106–189.

Waters, J. K. (2014). *The great adaptive learning experiment*. Retrieved November 2020, from https://campustechnology.com/Articles/2014/04/16/The-Great-Adaptive-Learning-Experiment.aspx

Zhu, Z., Sun, Y., & Riezebos, P. (2016). Introducing the smart education framework: Core elements for successful learning in a digital world. *International Journal of Smart Technology and Learning*, *1*(1), 53–66.

Zhu, Z. T., Yu, M. H., & Riezebos, P. (2016). A research framework of smart education. *Smart Learning Environments*, *3*(1), 1–17.

2 Intelligent Affect-Sensitive Tutoring Systems
An Evaluative Case Study and Future Directions

Hossein Sarrafzadeh
and Farhad Mehdipour

CONTENTS

DOI: 10.1201/9781003215349-2

2.1 INTRODUCTION

The human brain has been described as a system in which emotion and cognition are inextricably integrated (Cytowic, 1989). Systematic scientific exploration has yielded, and continues to yield, insights into the relationship between outwardly rational or logical cognitive processes and underlying affective state. The most immediately apparent effect that emotion has on cognitive events is on motivation. An individual's internal assessment, or feeling, about a situation will largely shape the action or behavior that is exhibited in response to a given situation. Thus, emotions are often considered as precursors of motivations (e.g., Oatley, 1992).

There is evidence of many roles of emotion in influencing aspects of cognitive performance and decision-making. Peters (2006) proposed a classification of some of these major roles. Among these, affect is considered to play a role as a source of information, to provide context and meaning to situations or events (Schwarz & Clore, 1988). Another role is that of a spotlight, directing the individual's attention on certain information and highlighting knowledge for further processing. This is similar to the concept of mood congruent memory (Bower & Forgas, 2001). Affect also functions as a motivator – this role is intuitively perhaps the most easily recognized. Affect influences the tendency to approach or avoid a situation as well as how information is processed (Frijda, 1986).

Emotion is thus strongly associated with whether information is attended to and how fast (or slowly) it is processed. Negative emotions are a widely agreed upon source of interference in the ability to process new information (Ellis & Ashbrook, 1988). It has been suggested that this may be an adaptational characteristic to encourage slower and more systematic processing. Negative emotions are a source of information to indicate that a problem has occurred and that future processing should be carried out in a more systematic and rational (and thus slower) manner (Schwarz, 1990). Certainly, it is apparent that affective state deeply influences many of the types of processes that take place during learning. As the learning experience is simultaneously something which is influenced by emotional state and also directly influences the emotional state of the learner – this relationship is both fundamental and complex.

These insights into the role that emotion plays gave rise to the proposition that a learning session may be improved if the teacher were sensitive and responsive to the emotional state of the learner. There exists a body of evidence to support this assertion in human to human tutoring dialogues (Goleman, 1995), and this interest has now been extended to computer-based learning (Alexander, Sarrafzadeh, & Fan, 2003; D'Mello & Graesser, 2012).

The central principle is generally similar: If a tutoring system were able to respond to the learners' emotional state, there would be measurably improved learning outcomes. A system that did this could incorporate insight into the learners' emotional state to guide the course of the lesson, for example, to offer hints when confusion is detected or to offer a break when boredom sets in. Research in the domain of affective tutoring systems (ATS) has yielded several exciting developments. These systems often incorporate functionality including tutoring strategies, affect sensing and learner progress tracking into a single, highly integrated environment. Empirical evaluations of these ATS support the fact that there can be generally positive results and measurable improvements in learning (e.g., Alexander, Sarrafzadeh, & Hill, 2008; Conati, 2002; D'Mello, Lehman, & Graesser, 2011). However, there is often no practical way to directly apply these findings to a real learning setting to discover the optimum environment for ATS success. As such there is no widespread use of ATS outside the controlled research setting.

This chapter provides a comprehensive review on the ATS and existing developments in the field and reports on the evaluation of the effectiveness of an ATS called Genetics with Jean, in which the animated tutor Jean responds to the affective state of the learner. This ATS is built using an existing affective application model (Thompson, 2012) and thus differs from existing systems due to the loose coupling between the tutoring and affect sensing functionality. This is advantageous as it provides direct support for reuse of software components outside the research environment and introduces the prospect of layering affect sensing functionality into existing, widespread e-learning environments. In the last section of the chapter, the results of evaluation and future directions and anticipated developments are discussed.

2.2   THE ROLE OF AFFECT IN LEARNING

Stein and Levine (1991) have proposed a goal-directed, problem-solving model which links a learner's goals and their emotional state. A principle to this model is that people prefer to be in certain states (e.g., pleasure) rather than others (e.g., pain): when unpleasant states are experienced, an attempt is made to regulate and to change these states toward more favorable ones by initiating the appropriate thinking or action. Stein and Levine propose that people continuously monitor their environment in this attempt to maximize positive states and when new information is detected; this interrupts the otherwise routine pattern matching process and attention may shift to the new information. Thus emotional experience is always associated with acquiring and processing new information, and learning always occurs during an emotional episode.

Another prominent theory linking emotional states with learning is the four quadrant model developed by Kort, Reilly, and Picard (2001). In this "learning spiral," the two axes signify learning and affect. The learning axis contains labels to indicate a range from constructive learning at one end to unlearning at the other. The affect axis ranges from negative to positive. Thus, proceeding in a spiral fashion, if a learner is working through an easy task, they will experience a combination of constructive learning and positive affect. If the material becomes more challenging, or they reach a troubling section, they may proceed to Quadrant II (constructive learning-negative affect) or Quadrant III (unlearning negative affect) before ultimately proceeding to Quadrant IV (unlearning-positive affect). At this point they may be uncertain how to progress, but as they reflect on the situation and process their existing knowledge they will eventually progress back to Quadrant I so that the spiral may continue as they acquire more knowledge.

The benefit of emotional awareness during human–human tutoring dialogues is already established.

Expert teachers are able to respond to the emotional states being experienced by students and guide their progress in a way that has a positive impact on learning (Goleman, 1995). While it has not been possible to precisely pinpoint how this is carried out, a fundamental act is the recognition of potentially detrimental negative states followed by guidance to a more positive state conducive to learning. The next logical step is to incorporate these findings into computer-based learning systems in an attempt to obtain the benefits of affect support.

The body of research into affective user interfaces has demonstrated that such timely provision of affect support does alleviate negative feelings (Klein, Moon, & Picard, 2002). These findings have been replicated and implemented in a number of environments including the use of animated agents (e.g., Prendinger, Dohi, Wang, Mayer, & Ishizuka, 2004) or tutoring systems (Sarrafzadeh, Fan, Dadgostar, Alexander, & Messom, 2004; Sarrafzadeh, Alexander, Dadgostar, Fan, & Bigdeli, 2008, Lin, Su, Chao, Hseih, & Tsai, 2016). These findings further support the proposition that the incorporation of affective components into tutoring systems – ATS - may result in enhanced learning outcomes.

2.3   AFFECTIVE TUTORING SYSTEMS

Research into the area of ATS is active and promising, the outcomes of which create new opportunities for human–computer interaction and are thus valuable both to educators and the wider community of developers. There is a number of ATS described in published literature with a few of the more prominent and successful projects being strongly represented. Each ATS tends to use a different set of input modalities and is tailored to work with a set of particular instructional material in a specific context. The internal architecture of these systems is often highly complex and not described in detail, simply because it is often not considered to be relevant when evaluating learning outcomes.

Although presented and constructed very differently, all ATS share certain characteristics: an ATS naturally requires some sort of affect sensing functionality (an affective model); there may also be a domain model to incorporate subject matter knowledge and/or a student model to incorporate the knowledge about the learner to guide the tutor to respond appropriately.

Various technologies have been used to implement each of these models. Conati et al. have used eye-tracking to gather input about the user's behavior, but they adapt hints in their game system without explicitly determining an emotional state (Conati et al., 2013). Sottilare and Proctor have focused on capturing user manipulations (e.g., mouse tracking) and using that to predict emotional state (Sottilare & Proctor, 2012). Tian, Gao, et al. use a restricted view of the user's content selection to determine emotional state (Tian, Gao, et al., 2014). High-quality cameras are now ubiquitous, and this makes it feasible for systems to remotely measure the user's physical characteristics. Several systems now use facial expression recognition algorithms to ascertain affective state (Sarrafzadeh, Hosseini, Fan, Overmyer, 2003; Baldassarri, Cerezo, Hupont, & Abadia, 2015; Lin, Wang, Chao, & Chien, 2012; Dadgostar, F., Sarrafzadeh, A, Overmyer, S., 2005).

There are also recent systems that focus primarily on the student model. These systems adapt their behavior based only upon the user's performance and hence cannot be truly considered "affective," but research in the field of affect dynamics suggests that there are direct correlations between performance and affect when using tutoring systems (Bosch & D'Mello, 2017). The Adaptive Peer Tutoring Assistant is one such system that supports the teaching of high school algebra (Walker, Rummel, & Koedinger, 2014); Ask-Elle is another example that teaches coding in the Haskell programming language (Gerdes, Heeren, Jeuring, & van Binsbergen, 2017).

Truly ATS will include elements from all three of the models. In spite of these high-level commonalities, due to the lack of standard implementation architecture, it is not possible to directly compare approaches aside from concluding that each ATS is either successful or unsuccessful in its own way. The following sections describe some of the most prominent ATS described in the literature.

AutoTutor is an intelligent tutoring system that interacts with learners using a natural language and helps them to construct explanations in simulated environments (Graesser, McDaniel, & Jackson, 2007). AutoTutor has gone through a number of iterations with different versions exploring various aspects of learning and modes of interaction. An affect sensitive version that was developed uses physiological and facial expression analysis and conversational cues to infer emotions being experienced by the learner. The emotions detected include boredom, engagement, confusion, and delight. The responses given by the tutoring system are subsequently selected to minimize the impact of any negative emotions. Initial results indicated that the affective tutor improved learning (as compared to a non-affective implementation of AutoTutor), particularly for learners with relatively low knowledge of the subject matter (D'Mello, et al., 2011). Internally, AutoTutor operates as a distributed client-server application with separate functional components handling speech act classification, user interface, conversation dialogue, and the generation of responses to the learner's conversational cues (D'Mello & Graesser, 2012). Central to this is the database of curriculum scripts, which contains the content associated with a question or exercise. The curriculum script must conform to certain constraints and include content such as ideal answer, potential hints, common misconceptions, keywords, and so on.

Prime Climb is an educational game developed at the University of British Columbia. The game is a multi-player climbing game which requires students to solve mathematical problems in order to proceed. Versions of the game have been created that incorporate an internal affective model to guide the action of an on-screen agent based on the most likely affective state of the learner (Conati & Zhao, 2004). This ATS monitors a user's emotions and engagement during their interaction with the game (Conati, 2002) and takes into account the users' physiological signals, as well as elements of user interface interaction to estimate the underlying emotional state based on a cognitive model of emotions (Ortony, Clore, & Collins, 1990). Both student's knowledge and emotional state can then be used to select appropriate actions to be delivered by the user interface (Hernández, Sucar, & Conati, 2008).

Easy with Eve is an affect sensitive mathematics tutor developed by the Next Generation Tutoring Systems project (Sarrafzadeh, Hosseini, Fan, Overmyer, 2003; Alexander et al., 2005; Sarrafzadeh, et al., 2008) at Massey University in New Zealand. This implementation utilizes observation to infer the learners affective state based on the facial expressions displayed. Video analysis is used to extract facial features during the learning session; these are then mapped to a set of "basic" emotions (Ekman & Friesen, 1978) by a fuzzy expression classifier. This information is subsequently utilized in guiding the behavior of the on-screen animated agent "Eve" with a case-based reasoning approach derived from observations of human tutors (Alexander, 2007). To develop this ATS the authors took the route of conducting a study on actual human tutors and performing a video analysis of tutoring dialogues. The findings were then encoded into rules and actions for the affective tutor to draw from using a case-based-reasoning approach. This method was believed to hold the greatest promise for success as it was modeled on actual human tutoring dialogues with expert teachers.

Almurshidi et al. have proposed a desktop-based intelligent tutoring system for helping students to learn about diabetes, different types and how it can be diagnosed. The system provides immediate and customized feedback to learners (Almurshidi & Abu Naser, 2017).

An adaptive learning interface has been introduced in Yuksel, et al. (2016) for learning piano that adjusts the task difficulty based on the learner's cognitive workload or brain state. Measurement of the cognitive workload is carried out using a brain-sensing technology so-called functional near-infrared spectroscopy (fNIRS). The proposed system called Brain Automated Chorales (BACh) helps users learn to play piano faster and with increased accuracy.

These examples are indicative of the breadth and diversity of development in this research domain, and there are many more diverse and unique applications. Each of these examples has several commonalities apart from their ability to detect and respond to emotions. Firstly, although implemented differently, the basic areas of functionality described at the start of this section are always present – this includes a student model, an affective model, and a domain model. Secondly, each of these ATS has been evaluated and has demonstrated some measurably improved learning outcomes. However, a third, and less fortunate area of commonality also exists and that is the fact that there is no direct way to replicate and extend these systems for a non-expert to create new lesson material to be taught by these existing ATS. This is an issue that has been identified by developers of some existing ATS and has resulted in developments such as the AutoTutor lesson authoring tool (Susarla, Adcock, Van Eck, Moreno, & Graesser, 2003); however, for the most part there does not yet exist a viable solution for all ATS platforms.

The Affective Stack Model (Thompson, 2012) holds promise to provide a solution to this issue. The Affective Stack Model provides an architecture in which the main functionality is described as a set of loosely coupled modules. The architecture is such that (a) individual functional components are not tied to a particular implementation and (b) the system is not tied to a particular instructional context. This translates into a system that supports rapid development, incremental improvement, and the ability to apply affective functionality to an existing tutoring (or other) software.

The use of a common architectural model such as the Affective Stack Model generally requires standardization of the functional components used in software. In this case, the "standardization" has arisen through observation, rather than by mandate. As described in Section 2.3, the common ATS do all include many of the same functional components, and these common components have therefore been incorporated into the Affective Stack Model. However, it is important to note that there is no perfect "one size fits all" approach. The ATS described in the literature do consist of complex and highly coupled functional components; however, it may sometimes be necessary to support the kind of complex and customized processing that takes place.

The success of these individual approaches is indeed evidenced by the research findings. Therefore, the central question is whether a more streamlined and simplified application based on the Affective Stack Model is able to still yield positive outcomes. If these are found, this will demonstrate the potential for existing tutoring software to be enhanced with affective capability. The following section describes the evaluation of an ATS that was built on top of an existing set of web-based instructional materials, using the Affective Stack Model as a development framework. The affective functionality was added to the regular software while simple text files were used for communication between components. As this is a prototype system, there is scope for any positive findings to be further enhanced in future with the development of a more complex and rich feature set.

2.4   GENETICS WITH JEAN

Genetics with Jean is an ATS developed in Western Australia that teaches the subject of genetics. Information is presented in a number of modalities including text, graphics, diagrams, and animation which are all based on the Morgan Genetics Tutorial (Sofer & Gribbin, 2017). The learner's affective state is inferred by physiological signals using a previously developed affective platform (Thompson, Koziniec, & McGill, 2012). The underlying affective model uses a dimensional view of emotions, in which the affective state is considered in terms of the components of affective valence and activation. The physiological signals measured were selected to give insight into these two dimensions and to thus classify the likely affective state of the learner. The system is self-calibrating and signals are considered in the context of the learners own recent range of physiological expression rather than comparison against a pre-selected "baseline." The insight into the affective state is then used to guide the actions of the on-screen animated agent, Jean. The purpose of the animated agent is to provide guidance and support to the user and to emulate a human tutor (Alexander, Hill and Sarrafzadeh 2005; Sarrafzadeh, Gholam Hosseini and Fan, 2003). The agent was based on the Microsoft Agent environment (Microsoft Corporation, 2009). The character has many animations, which may be scripted within software to achieve a believable and natural interaction. In order to make the character appear more natural, speech balloons were accompanied by a spoken voice. This served the purpose of making the agent more believable as a learning companion.

As noted previously, ATS are often highly complex and require specialized software artifacts by using precisely calibrated conditions and equipment. This ultimately limits the widespread applicability of the findings. Therefore, in this implementation, care was taken to utilize technology which may be conducive for later transference into consumer applications.

Two physiological sensors were utilized to infer the dimensions of affective activation (the extent of activation or arousal experienced by the individual) and valence (how strongly positive or negative this experience is). The sensors consist of small (3 mm) electrodes which simply require skin contact to operate. These were affixed to the skin with adhesive tape during the evaluation; however, it is anticipated that in future refinements to the hardware these may be embedded into existing interface mechanisms such as keyboard or mouse. This concept has been previously demonstrated (e.g., Kirsch, 1997), and it is hoped that it will be developed further as the technology improves.

The emotional activation of the learner is inferred by measuring the conductivity of the skin. Skin conductivity tends to increase when a person is startled or experiences anxiety and is generally considered to be a good measure of a person's overall level of activation (Cacioppo, Tassinary, & Berntson, 2007). Tonic (background level) skin conductance varies with psychological arousal, rising sharply when the subject awakens and rising further with activity, mental effort, or especially stress (Woodworth & Schlosberg, 1954).

The emotional valence of the learner can also be inferred using heart rate-based measures. While these are commonly measured using an electrocardiogram using electrodes on a chest strap, this is quite an intrusive means of physiological measurement and this detracts from its usefulness, especially for an affective computing application (D'Mello, 2008). Therefore, in this implementation a more suitable approach was developed which simply requires skin contact using the same 3 mm sensors described above. Heart rate variability measures provide a rich source of information and make it possible to analyze specific frequency bands which correspond to certain underlying processes, for example, to discern between physiological changes due to physical exertion as opposed to affective state (Yannakakis, Hallam, & Lund, 2008). Therefore, in this implementation heart rate variability was chosen as the measure of emotional valence.

The ATS operates in a fully automated real-time processing mode. This is a mandatory requirement in order for the system to function in its intended role as a real-time, interactive tutor. In keeping with the aim of promoting transferability, the reliance on commercial hardware platforms was kept minimal, with a generic National Instruments data acquisition device used to collect the physiological signals. This approach also made it possible to create a physiological data acquisition system that was superior in many ways to the commercially available products. This is particularly noteworthy in the implementation of heart rate variability measures. It was found that the prominent commercially available platforms such as those available from ProComp and BioPac (Strauss, et al., 2005) often carry certain limitations, the most apparent being the choice of physical sensors and form factors, as these devices are designed only to operate with the particular choice of hardware sensors that are provided by the manufacturer. A review of the current offerings from ProComp and BioPac also revealed that the analysis of heart rate variability measures was often only supported in off-line processing mode – a fact which alone renders them unsuitable for a real-time application such as an affective tutor. A further improvement over commercial platforms was in the responsivity of the system. Heart rate variability measures operate using blocks of data – the size of the block will influence how responsive the system is to detecting changes. Commercial platforms generally offer fixed pre-sets for data block size such as 3, 5, or 10 minutes (Thought Technology 2010) or allow the user to pre-select a block in offline mode (BioPac Systems 2004). In this implementation, signal pre-processing techniques were developed which make it possible to operate successfully using a much smaller block size of 100 seconds of data. This translates to a system that is more responsive to changes in the users underlying affective state.

In this software, the animated agent serves as an affective tutor, which aims to address, and alleviate, negative affective states so that they do not hinder the learning process. A number of scripted behaviors were incorporated into this application and these fall into three distinct modes described as follows:

Support Affect: The agent provides empathic feedback for the user to address any potentially negative affective state; for example, *"I know the material is quite hard"* or *"It's ok if you don't score full marks, you can always come back to this another time."*

Motivate: The agent provides supportive comments to keep the learner on track; this may simply be *"Well done,"* or *"Keep up the good work."* If the lesson is nearing the end, then the agent may also indicate that the task is almost complete if the user is losing interest.

Offer Revision: The agent provides extra tutoring for topics that have triggered a negative affective state in the user or when the user's performance on quizzes is poor. This may take the form of separate revision pages with a summary of a whole topic or simply links to enable the user to easily look back to re-check content on a previous page.

The classification of affective state based on activation/valence dimensions was based on that of Prendinger et al. (2004) and the following general rules were applied:

1. **No action** taken if affective state is neutral.
2. **Offer Revision** triggered if lesson performance is low.
3. **Support Affect** triggered if negative state detected (this is characterized by higher activation and negative valence).
4. **Motivate** response given if high activation detected with positive valence. This could indicate frustration or simply distraction. This response is generally motivational and does not address any specific state. These rules are summarized in the decision tree as shown in Figure 2.1.

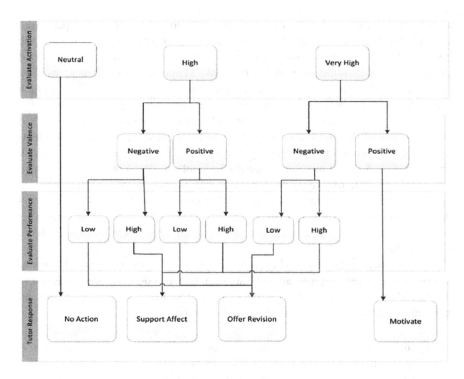

FIGURE 2.1 Classification of states and actions.

2.5 EVALUATION OF THE GENETICS WITH JEAN ATS

An empirical evaluation was carried out to determine the effectiveness of Genetics with Jean in improving outcomes associated with use of an ATS. The evaluation focused on the contribution of affective support to the effectiveness of the system.

Effectiveness, in an e-learning usability context, is broadly defined as the attainment of instructional objectives or that the learning outcomes are consistent with the learners' expectations (Pajares, 1996). For the purposes of this study, effectiveness is considered in terms of three dimensions: Content Knowledge, Perceived Learning, and Enjoyment. Content Knowledge is defined as the extent to which facts about the lesson content are retained by the learner. Perceived learning is defined as the self-reported perceptions of learning accomplishments (Jiang, 2000). Enjoyment refers to the extent to which the activity of using the system is perceived to be enjoyable in its own right, apart from any performance considerations that may be anticipated (Malone, 1981) and is thus defined as a form of intrinsic motivation (Davis, Bagozzi, & Warshaw, 1992).

Lisetti (1999) claims that a large number of cognitive tasks are influenced by affective state, including organization of memory, attention, perception, and learning. The same conclusion was reached by Picard (1997, p. 28) who states that "emotions play an essential role in rational decision making, perception, learning and a variety of other cognitive functions." Therefore, students whose learning is supported by this affective e-learning software could be expected to retain more knowledge of the content than students who do not receive this support. This leads to the first hypothesis that

> *H1: Students who receive affective support from Genetics with Jean will have higher levels of Content Knowledge than students who do not receive affective support.*

In practice, learning a new skill or knowledge is a complex process, characterized by many internal processes, which may not be immediately observable or measurable. From a cognitive perspective, learning is considered in terms of changes in mental models or knowledge representations (Shuell, 1981). Thus, while these changes may indeed be occurring, task performance is only one possible outcome of this process. According to Ausubel (1968), successful performance requires not only knowledge acquisition (learning) but also other factors such as perseverance, flexibility, or improvisation. Thus, performance is not the only indication that learning has occurred. Therefore, students whose learning is supported by affective e-learning software could be expected to have a higher level of perceived learning than students who do not receive this support. This leads to the second hypothesis that

> *H2: Students who receive affective support from Genetics with Jean will have higher levels of Perceived Learning than students who do not receive affective support.*

Increased enjoyment and engagement have implications for learning. Csíkszentmihályi (1990, 2014) described an ideal learning state, which he called the zone of flow. In this state, time and fatigue disappear as the learner is absorbed and immersed in the task they are undertaking. When in a state of flow, people are absorbed in the activity and feel in control of the task and environment (Hsu & Lu, 2004). Thus, the engagement and enjoyment of the learner is a catalyst to mediate their future learning and interest (Fu, Su, & Yu, 2009).

An ATS responds to the learner's affective state and attempts to provide feedback and support to address any negative states that may arise. Therefore, students who receive the feedback and support from an ATS could be expected to find the lesson more enjoyable than those who did not receive this feedback and support. This leads to the third hypothesis that

> *H3: Students who receive affective support from Genetics with Jean will have higher levels of Enjoyment than students who do not receive affective support.*

2.5.1 METHODOLOGY

Data was gathered during a laboratory experiment in which participants interacted with the Genetics with Jean ATS. To achieve the objective of evaluating whether the fully functional Genetics with Jean ATS would be more effective than a non-affective version, participants were randomly assigned into two groups. The affective group interacted with the fully functional ATS, whereas for the non-affective group, affective support was disabled.

2.5.1.1 Development of Measurement Instruments

As discussed above, the overall effectiveness of Genetics with Jean was considered in terms of three dimensions: Content Knowledge, Perceived Learning, and Enjoyment. This section discusses the development of the instruments used to measure the aspects of effectiveness considered in the study.

2.5.1.2 Content Knowledge

Content Knowledge was measured using a summary quiz conducted at the end of the data-collection session. The summary quiz consisted of 12 questions which were based on the content covered in the lesson and worth one mark each. The questions were a mixture of fill in the blank and true/false questions. The total quiz score for each participant was calculated by adding the individual marks giving a maximum attainable quiz score of 12. The quiz questions are detailed in Table 2.1.

2.5.1.3 Perceived Learning

There are aspects of the learning experience that may not necessarily be reflected in test scores. Perceived Learning was measured using five items, adapted from Alavi, Marakas, and Yoo (2002). These items were measured on a five-point Likert scale, ranging from strongly disagree (1) to strongly agree (5). An overall score for the Perceived Learning was calculated as the average of the five items. The items are shown in Table 2.2.

TABLE 2.1
Content Knowledge quiz questions

DNA consists of ___ of genes arranged in a specific order.

The genes which carry the sequence of particular proteins is specified in a stretch of ___.

The long pieces of DNA are called chromosomes. T/F

Human beings have tens of thousands of different kinds of proteins, and each one has a specific function. T/F

Each species has the same number of chromosomes. T/F

During meiosis the number of chromosomes gets reduced by a factor of ___.

The choosing of chromosomes to be passed on during meiosis is a random process. T/F

Sex cells or gametes contain half of the number of chromosomes of other cells. T/F

A die is rolled, find the probability that an even number is obtained. ___ out of ___.

Two coins are tossed, the probability that two heads are obtained. ___ out of ___.

A card is drawn at random from a deck of 52 cards. Find the probability of getting a queen. ___ out of ___.

A jar contains 3 red marbles, 7 green marbles, and 10 white marbles. If a marble is drawn from the jar at random, what is the probability that this marble is white? ___ out of ___.

TABLE 2.2

Perceived Learning items

I became more interested in the subject of genetics

I gained a good understanding of the subject of genetics

I learned to identify central areas in the subject

I was stimulated to do additional study in genetics

I found the current lesson to be a good learning experience

2.5.1.4 Enjoyment

Enjoyment was measured using four items on a five-point semantic differential scale. The items were adapted from Ghani and Deshpande (1994). Details of the items are included in Table 2.3.

2.5.2 PARTICIPANTS

The participants for this experiment were recruited from staff and students of two Australian universities. The primary selection criteria were that participants must be in good overall health and not be under the influence of any medication or drugs (including caffeine) which may affect their physiological state. An additional criterion was that participants must not already possess a high level of knowledge in the subject matter taught by the Genetics with Jean ATS.

Potential participants were invited to take part in the study via email distribution lists or referred by previous participants. All participants were sent a copy of the information letter so that they would fully understand the requirements before committing to taking part. Prior to commencement of the recruiting process, ethics approval was sought and obtained.

2.5.3 DATA COLLECTION SESSION

Separate data collection sessions were carried out for each participant. These were held in a quiet computer laboratory with only the participant and researcher present to minimize any distractions. The physiological sensors were attached to two fingers of the non-dominant hand, to ensure that their presence did not interfere as the participants interacted with the software. As all participants exhibit a slightly different

---

TABLE 2.3

Enjoyment items

I found the lesson:

Interesting .... Not interesting

Fun .... Not fun

Exciting .... Not exciting

Enjoyable .... Not enjoyable

---

range of physiological responses, the Genetics with Jean ATS performs a self-calibration during operation. To ensure that this was completed, the sensors were attached at least two minutes before the start of the tasks to ensure that sufficient data had been recorded for the software to reliably calibrate.

Participants were randomly assigned to either the affective group or the non-affective group using a software randomizer – the affective group interacted with the fully functional ATS, the non-affective group used identical software, but with the physiological input disabled. The physiological sensors were attached in the same way to participants in both groups to ensure that they were treated consistently.

The use of the Affective Stack Model to guide the development of the Genetics with Jean ATS meant that system components were not necessarily dependent on one another. Therefore, the ATS could operate correctly even in the absence of any physiological input. This characteristic streamlined the way in which the experiment was conducted as it meant that all participants could be treated in the same way and run the same software.

The Genetics with Jean ATS interacts with the user via an on-screen animated agent intended to emulate a human tutor, Jean. In order to be able to ensure that any observed differences could be attributed to the affective components of the software and not simply the presence of an animated agent, the animated agent was present for both groups. In the version of the software used by the non-affective group, the animated agent was unable to detect affective cues and thus relied only on non-affective sources of input. This was limited to reporting on the student's progress in the lesson and presenting revision topics if the student requested help.

The sequence of steps for the study and details of the activities undertaken in the data collection sessions are shown in Table 2.4.

The study procedure was explained during Part 1. Participants went through the project information letter and consent form and were able to ask any questions about the tasks and clarify anything they wished. After this, the measuring equipment was set up and the sensors were attached. The measuring equipment to be used in the experiment was left running during the remainder of the introductory discussion and software demonstration to allow the participants to become familiar with the setup.

After the introductory discussion, the software was demonstrated to the participant to ensure that they were aware of the navigation controls and the general screen layout to expect in the lesson. While this was mostly self-explanatory, a demonstration was conducted so that all participants would be more comfortable when using the software

TABLE 2.4
Activities undertaken during the data collection session

Part activities	Approx. duration
1 Introduce study and obtain participant's consent	10 min
Demonstrate software	
2 Use ATS	25 min
3 Questionnaire	10 min

interface and to ensure that ease of use (or lack of) did not influence the user's perceptions of the lesson itself.

Part 2 of the study required the user to interact with the ATS to complete several short lessons and quiz questions on the subject of genetics. In the affective group, the behavior of the on-screen animated agent was determined by the affective responses displayed by the participant. The ATS was able to sense and respond to any negative affective states being experienced by the participants.

The final stage of the data collection session, Part 3, comprised completion of a questionnaire. The questionnaire was filled out electronically and covered the three aspects of effectiveness: Content Knowledge, Perceived Learning, and Enjoyment.

The sections in the questionnaire were ordered as follows and participants were also given a blank text area in which they could include any other feedback or observations about either their Perceived Learning or Enjoyment of the lesson:

1. Age and gender
2. Perceived Learning
3. Enjoyment
4. Content Knowledge quiz.

The Perceived Learning and Enjoyment items involve the participant's self-report of their internal state or experiences. Therefore, it was desirable to present these items as soon after the lesson as possible while the experiences were fresh in their memory. Furthermore, this presentation order also precluded the possibility of any negative affect experienced when answering the Content Knowledge quiz from interfering with the measurement of Enjoyment or Perceived Learning. An additional benefit of putting the quiz session at the very end was that it may have made the test slightly more challenging, as the material may not be as easy to recall after having to attend to another secondary task.

2.5.4 PILOT

Data sets from the first five participants were given additional scrutiny involving visualization and manual pattern analysis, and this was compared with the data logs generated by the software. The reason for this was to provide an additional level of testing and to give an early warning if there were any issues with the hardware and software. This analysis confirmed that the Genetics with Jean ATS was operating correctly, and the data collection could proceed.

A total of 40 participants took part in the study and these were equally distributed into the two groups. The sample consisted of 22 female and 18 male participants with the age of over half of the participants (55.0%) being in the range of 20 to 29 years, and the second most common age group (22.5%) being 30 to 39 years.

Reliability testing was conducted to ensure that the items used to measure Perceived Learning and Enjoyment demonstrated sufficient internal consistency. Cronbach's alpha for Perceived Learning was 0.7, and the scale was thus found to be reliable (Nunnally, 1978). Cronbach's alpha for Enjoyment was 0.67. While this is slightly lower than the desired threshold this was deemed to be acceptable for an exploratory study.

2.5.5 Does Affective Support Lead to Improvements in Students' Knowledge?

A difference in summary quiz score between the affective group and the non-affective group would suggest that affective support provided by the Genetics with Jean ATS leads to improvements in the students' knowledge of the lesson content. Descriptive statistics for the two groups are presented in Table 2.5. Content Knowledge scores were distributed normally, and independent samples t-tests were considered suitable to test for differences between the two groups. The results of the independent samples t-test indicated that there was no significant difference in levels of Content Knowledge between the affective (M=10.60, SD=1.23) and non-affective group (M=10.25, SD=1.74); $t(38)$= -0.73, p= 0.47. Therefore, the hypothesis that affective support increases acquisition of knowledge about the subject area, genetics, was not supported.

2.5.6 Does Affective Support Lead to Improvements in Students' Perceived Learning?

Descriptive statistics for the affective and non-affective groups are presented in Table 2.6. The Perceived Learning data did not meet the assumption of normality required for an independent samples t-test. Therefore, a Mann–Whitney U test was conducted to evaluate the hypothesis that students who receive affective support from the Genetics with Jean ATS will have higher levels of Perceived Learning than students who did not receive affective support.

Participants in the affective group reported greater levels of Perceived Learning than those in the non-affective group (Mdn 4.00 vs 3.40; U= 94.0, Z= -2.90, p= 0.004) and the hypothesis was thus supported.

TABLE 2.5
Content Knowledge group statistics

Group	N	Min	Max	Mean	SD
Non-affective	20	6.0	12.0	10.25	1.74
Affective	20	8.0	1g2.0	10.60	1.2

TABLE 2.6
Perceived Learning group statistics

Group	N	Min	Max	Mdn	SD
Non-affective	20	2.60	4.40	3.40	0.40
Affective	20	2.80	5.00	4.00	0.50

TABLE 2.7
Enjoyment group statistics

Group	N	Min	Max	Mdn	SD
Non-affective	20	3.00	4.50	3.75	0.47
Affective	20	3.00	5.00	4.00	0.68

2.5.7 DOES AFFECTIVE SUPPORT LEAD TO IMPROVEMENTS IN STUDENTS' ENJOYMENT OF LEARNING?

Descriptive statistics for Enjoyment are shown in Table 2.7. The enjoyment data did not meet the assumption of normality required for an independent samples t-test, therefore a Mann-Whitney U test was conducted to evaluate the hypothesis that students who received affective support as part of the Genetics with Jean ATS would have higher levels of Enjoyment than students who did not receive affective support. This hypothesis was not however supported with the results indicating that there was no significant difference in the levels of Enjoyment between the affective and non-affective groups (Mdn 4.00 vs 3.75; $U= 148.50$, $Z= -1.40$, $p= 0.16$).

2.6 DISCUSSION

In this work, we evaluated the efficiency of an ATS called Genetics with Jean. The system teaches introductory genetics, interacting with the user via an animated agent which has spoken and on-screen dialogue. An internal decision network utilizes data from the affective platform to infer the dimensions of affective activation and valence in real-time while the learner interacts with the system. It has the capability to perform various types of interaction including providing affect support to address negative states, encouraging the learner and providing revision and review tips.

This ATS was developed to address limitations of previous ATS: firstly that there is often no practical way to implement these systems in real-world settings for long-term use; and secondly that it has not been easy to add affect support to existing tutoring systems. Genetics with Jean differs from previous ATS in that it used the open and re-usable Affective Stack Model (Thompson, 2012) as a framework for development. The approach taken provides a way forward from the current ad hoc nature of ATS development.

The use of a component-based model brings together all the functional units of an affective computing environment in an architecture that is compatible with third party software applications, enables developers to incrementally improve separate components, and supports the reuse of functional components for new applications. Its use in the development of Genetics with Jean resulted in loose coupling between the tutoring and affect sensing functionality, which is advantageous as it provides direct support for reuse of software components outside the research environment and introduces the prospect of layering affect sensing functionality into existing, widespread e-learning environments.

The effectiveness of Genetics with Jean in terms of supporting student learning was evaluated in terms of three dimensions: Content Knowledge, Perceived Learning, and Enjoyment. Although the means for all three dimensions of effectiveness were higher for those students receiving affective support from the ATS than for those who did not receive it, these differences were only significant for Perceived Learning. This finding for Perceived Learning is consistent with that of Alexander (2007) who noted that students who interacted with his affective tutor had marginally higher levels of Perceived Learning than those who interacted with a non-affect sensing version.

The lack of significant improvement in Content Knowledge, as measured by a summary quiz, might result from the levels of existing domain knowledge of the participants. The means of both the affective group and the non-affective group were quite high compared with the maximum attainable score of 12, suggesting that a reason for the lack of significant difference might be found in a ceiling effect whereby the high scores obscured differences in group performance. It is possible that more challenging material or a different assessment technique would not have this ceiling effect and thus any differences between groups would be clearer. This is consistent with the findings of D'Mello, et al. (2011) who note that the affective implementation of their AutoTutor software is particularly appropriate for low domain knowledge learners. Aist, et al. (2002) also made a similar observation that including affective enhancements in a tutoring system improved students' persistence with the task, but not their memory of facts. This implies that the benefits may be more apparent in the longer term rather than after a short evaluation. This should be investigated in future research.

Levels of Enjoyment were also not found to significantly differ in response to the affective support. The reason for this may lie in the nature of enjoyment of learning. Ghani (1991, 1994) noted that the dimensions of task challenge and sense of being in control were key factors that resulted in being in a state of enjoyment. The role of these factors may also explain the results with respect to Enjoyment in this study. Both groups of learners interacted with an identical set of materials within the tutoring system and hence should have experienced a similar level of sense of control and task challenge; explaining the similar levels of Enjoyment. However, given that the median Enjoyment score was higher for the affective group, further research into this issue may be warranted. Another factor potentially indicating Enjoyment is the time spent on task. While this was not explicitly measured during the study, it was noted that almost all participants chose to interact with the ATS for longer than the anticipated study duration. Indeed, some participants spent a full hour using the software – far more than the 25 minutes originally allocated to that part of the session. Further support that the participants found the session enjoyable comes from the open-ended comments that some provided. One particular comment clearly describes the kind of flow experience that is sought in human–computer interaction: *"I'm tired out after that lesson, but I didn't realize I'd been concentrating on it for 45 minutes!"*. It is possible that this immersion and lack of awareness of time contributed to the observed long session durations. Prior studies have shown that time on task is a predictor of computer-based learning success (Brown, 2001), and thus this is a positive observation that may be useful to guide future developments.

Genetics with Jean is teaching material of a scientific nature. Lin et al. have reported success using a similar approach when the content area is digital arts (Lin et al., 2014, Hsu et al., 2014). The fact that ATS can work with such diverse subject matter is another positive observation about future developments.

2.7   CONCLUSION AND FUTURE DIRECTIONS

Emotion is recognized as an important factor influencing learning. ATS attempt to improve learning by providing affect support to students as part of the learning experience. The study described in this chapter involved the development and evaluation of an ATS called Genetics with Jean. This system was developed with the goal of being proof of concept for a development approach that should facilitate the enhancement of existing e-learning systems to provide affective support. Genetics with Jean was shown to improve students' perceptions of their learning, but not knowledge of the content as measured by a quiz. The results support the previous observations that affective tutors are perhaps most useful in situations with low-domain knowledge learners (D'Mello, et al., 2011).

Increased pressures and time-constraints on individuals' work and personal life require that teaching and learning techniques become more flexible, effective, and efficient to remain viable. The consistent growth in adoption of electronic learning, despite the evidence that it is often less effective than human tutoring (Bloom, 1984), demonstrates that the flexibility and ease of access are potentially the deciding factors. Therefore, advanced technologies such as ATS which combine the ease of access and flexibility of electronic learning with the effectiveness of human tutoring hold promise and are consequently attracting attention within the industry (Lowendahl, 2017). However, unless the field progresses to a point where it is possible to rapidly develop and reuse affective technological solutions, widespread uptake may potentially never occur.

Developments in affective computing applications are generally highly implementation specific. As affective applications are considerably specialized and complex, there has to date been little discussion regarding the concept of repurposing these applications. Although a great deal of time may be invested in the development of an affective interface for an electronic learning software, the complexity of later refining this application or porting it to a new implementation domain may ultimately be prohibitive. Baker (2016) suggests that this might account for the relative simplicity/narrowness of existing systems. Still, the development approach of utilizing a standard model of core functional components has shown potential as a way forward for future work in the domain of ATS. Harley et al. (2015) support this position via a meta-study that helps to unify the field of ATS research; they even develop a taxonomy to further discussion in the area. This chapter has detailed the evaluation of one such ATS that was developed using a standardized and conceptually straightforward architecture. The positive findings and areas for future investigation have laid the foundations for a promising direction for research and practice.

Sensor technologies play a vital role in the development of intelligent tutoring systems. There have been significant advances in digital biomarkers for measuring psychological features such as blood volume pulse (BVP) and electrodermal activity

(EDA). These devices could be used for characterizing major depressive disorder (MDD). Empatica E4 (Meyer, et al., 2020) is an example of a recently developed digital biomarker which is a wristband sensor that measures BVP and EDA and is helpful to assess psychological depressive symptom severity for individuals. According to several studies, some aspects of the introduced devices / technologies may be well received, but other aspects require further improvement in the comfort level and ease of use before utilizing them in real-world implementations.

Nowadays, a wide variety of e-learning platforms and tools are available while recently more attention is being paid to learner's needs and abilities by making the learning individualized. The development of affective ITSs, although progressing, it is not yet at the level to automatically adjust the learner's knowledge level and preferences. There are several challenges for fully realizing adaptive training due to various reasons. By using an adaptive system, it is possible to personalize the learning and adjust the content based on learner's interest; however, important parts of learning content may be omitted in this case. Adjusting learning content to learner's need is another issue that is addressed by researchers in different ways, where some attempted to standardize learning content, and some others are directed to adapting learning content to learner's preferences and goals. Relying on a single mechanism for detecting emotions may result in failure in identifying the correct affective state. With a multimodal mechanism it may be possible to overcome this issue. However, a question which arises is "what are the emotions of interest in the context of e-learning"? (Feidakis, 2016).

Capturing signals corresponding to various behavioral states from the scalp surface area through Electroencephalography (EEG) can help analyze human brain and its cognitive behavior. For a comprehensive analysis of emotional states of learners, psychological (e.g., blood pressure) and behavioral modalities (such as facial recognition, EEG signal analysis, and natural language recognition, and gesture recognition) could be combined. In the future, the role of personal devices in recognizing emotional states and creating personalized experiences is undeniable. Due to the popularity of virtual personal assistants (e.g., Google Assistant), conversation systems will further engage affective computing to understand and respond to users' emotional states. Real-time emotion analysis as a part of systems requires visual sensors, emotion-sensing wearables, and Artificial Intelligence (A.I.)-based emotion tracking to detect a user's mood and make necessary adjustments in the system. There is need for creation of generic and accessible platforms in the future rather than currently proprietary technologies built for certain use cases. Although there are already applications for reading body gestures and analyzing some psychological signals, these applications are comparatively less developed (Garcia-Garcia et al., 2018).

Emotions are vital part of humans' lives; hence, the ethical implications of affective computing should be appropriately addressed. Some of the ethical issues associated with affective computing are not completely new. Affective computing may pose risks to the body particularly when it comes to the uses of invasive technologies. Also, as all these systems collect sensitive data from individuals, the issues of data privacy and security need to be addressed. There are more risks associated with the affective systems when emotional states of individuals are directly manipulated which may cause irreversible consequences. To reduce ethical implications, it is important that people do have a clear grasp of what the technology does to them (Steinert & Friedrich, 2020).

REFERENCES

Aist, G., Kort, B., Reilly, R., Mostow, J., & Picard, R. (2002). *Experimentally augmenting an intelligent tutoring system with human-supplied capabilities: Adding human-provided emotional scaffolding to an automated reading tutor that listens.* Proceedings of the 4th IEEE International Conference on Multimodal Interfaces, 483–490. Pittsburgh, PA, USA.

Alavi, M., Marakas, G., & Yoo, Y. (2002). A comparative study of distributed learning environments on learning outcomes. *Information Systems Research, 13*(4), 404–415.

Alexander, S. (2007). *An affect-sensitive intelligent tutoring system with an animated pedagogical agent that adapts to human emotion.* Unpublished PhD Thesis, Massey University, Albany, New Zealand.

Alexander, S. T. V., Hill, S., & Sarrafzadeh, A. (2005, July). *How do human tutors adapt to affective state.* In *Proceedings of User Modelling.*

Alexander, S., Sarrafzadeh, A. (2004). Interfaces that adapt like humans, *Computer Human Interaction,* (pp. 641–645), Berlin/Heidelberg: Springer.

Alexander, S., Sarrafzadeh, A., & Fan, C. (2003). *Pay attention! The computer is watching.* In *Affective tutoring systems proceedings E-learn: World conference on E-learning in corporate, government, healthcare, and higher education,* Phoenix, AZ.

Alexander, S., Sarrafzadeh, A., & Hill, S. (2008). Foundation of an affective tutoring system: learning how human tutors adapt to student emotion, *International Journal of Intelligent Systems Technologies and Applications,* 4(3/4), 355.

Almurshidi, S. H., & Abu Naser, S. S. (2017). Design and Development of Diabetes Intelligent Tutoring System. *European Academic Research,* 6(9), 8117–8128.

Ausubel, D. P. (1968). *Educational psychology: A cognitive view.* New York: Holt, Rinehart and Winston.

BioPac Systems (2004). Heart rate variability analysis. Retrieved September 1, 2020, from http://www.biopac.com/Curriculum/pdf/h32.pdf.

Baker, R. S. 2016. Missing title. *International Journal of Artificial Intelligence Education,* 20, 600.

Baldassarri, S., Cerezo, E., Hupont, I., & Abadia, D. (2015) Affective-aware tutoring platform for interactive digital television. *Multimedia Tools and Applications,* 74(9), 3183–3206.

Bloom, B. S. (1984). The 2 sigma problem: The search for methods of group instruction as effective as one-to-one tutoring. *Educational Researcher, 13*(6), 4–16.

Bosch, N., & D'Mello, S (2017). The affective experience of novice computer programmers. *IJAIED,* 27(1), 181–206.

Bower, G. H., & Forgas, J. P. (2001). Mood and social memory. In J. P. Forgas (Ed.), *Handbook of affect and social cognition* (pp. 95–120). Oxford: Pergamon.

Brown, K. G. (2001). Using computers to deliver training: Which employees learn and why? *Personnel Psychology, 54*(2), 271–296.

Cacioppo, J. T., Tassinary, L. G., & Berntson, G. G. (Eds.). (2007). *Handbook of psychophysiology* (3rd ed.). New York: Cambridge University Press.

Conati, C. (2002). Probabilistic assessment of user's emotions in educational games. *Applied Artificial Intelligence, 16*(7–8), 555–575.

Conati, C., Jaques, N., & Muir, M, (2013) Understanding attention to adaptive hints in educational games: An eye-tracking study. *IJAIED,* 23("Best of ITS 2012"), 136–161.

Conati, C., & Zhao, X. (2004). *Building and evaluating an intelligent pedagogical agent to improve the effectiveness of an educational game. Proceedings of the 9th International Conference on Intelligent User Interfaces*, 6–13. Funchal, Madeira, Portugal.

Csíkszentmihályi, M. (1990). *Flow: The psychology of optimal experience.* New York: Harper and Row.

Cytowic, R. E. (1989). *Synesthesia: A union of the senses.* New York: Springer-Verlag.

D'Mello, S. (2008). Automatic detection of learner's affect from conversational cues. *User Modeling and User-Adapted Interaction, 18*(1–2), 45–80.

D'Mello, S., & Graesser, A. (2012). AutoTutor and affective autotutor Learning by talking with cognitively and emotionally intelligent computers that talk back. *ACM Transactions on Interactive Intelligent Systems, 2*(4), 1–39.

D'Mello, S., Lehman, B., & Graesser, A. (2011). A motivationally supportive affect-sensitive AutoTutor. In R. A. Calvo & S. K. D'Mello (Eds.), *New perspectives on affect and learning technologies* (Vol. 3, pp. 113–126). New York: Springer.

Dadgostar, F., Sarrafzadeh, A, Overmyer, S. (2005). *Face tracking using mean-shift algorithm: A fuzzy approach for boundary detection, Proceedings of International Conference on Affective Computing and Intelligent Interaction*, Springer, Berlin, Heidelberg, pp. 56–63.

Davis, F. D., Bagozzi, R. P., & Warshaw, P. R. (1992). Extrinsic and intrinsic motivation to use computers in the workplace. *Journal of Applied Social Psychology, 22*(14), 1111–1132.

Ekman, P., & Friesen, W. (1978). *Facial action coding system: A technique for the measurement of facial movement.* Palo Alto, CA: Consulting Psychologists Press.

Ellis, H. C., Ashbrook, P. W. (1988). Resource allocation model of the effects of depressed mood states on memory. In: K. Fiedler J. Forgas (Eds.), *Affect, cognition and social behaviour* (pp. 25–43). Hogrefe; Toronto.

Feidakis, M. (2016). A review of emotion-aware systems for e-learning in virtual environments. *Formative Assessment, Learning Data Analytics and Gamification*, 217–242. doi:10.1016/b978-0-12-803637-2.00011-7

Frijda, N. H. (1986). *The emotions.* Cambridge: Cambridge University Press.

Fu, F., Su, R., & Yu, S. (2009). EGameFlow: A scale to measure learners' enjoyment of e-learning games. *Computers and Education, 52*(1), 101–112.

Garcia-Garcia, J. M., Penichet, V. M., Lozano, M. D., Garrido, J. E., & Law, E. L. (2018). Multimodal affective computing to enhance the user experience of educational software applications. *Mobile Information Systems*, 2018, 1–10. doi:10.1155/2018/8751426

Gerdes, A., Heeren, B., Jeuring, J., & van Binsbergen, L.T. (2017). Ask-Elle: An adaptable programming tutor for Haskell, giving automated feedback. *IJAIED, 27*(1), 65–100.

Ghani, J. (1991). Flow in human-computer interactions: Test of a model. In J. Carey (Ed.), *Human factors in management information systems: An organizational perspective* (Vol. 3). Norwood, NJ: Ablex.

Ghani, J. & Deshpande, S. (1994). Task characteristics and the experience of optimal flow in human computer interaction. *The Journal of Psychology, 128*(4), 381–391.

Goleman, D. (1995). *Emotional intelligence.* New York: Bantam Books.

Graesser, A. C., McDaniel, B., & Jackson, G. T. (2007). Autotutor holds conversations with learners that are responsive to their cognitive and emotional states. *Educational Technology, 47*, 19–22.

Harley, J. M., Lajoie, S. P., Frasson, C., & Hall, N. C. (2015). *An Integrated Emotion-Aware Framework for Intelligent Tutoring Systems. Artificial Intelligence in Education. AIED 2015. Lecture Notes in Computer Science*, vol 9112. Springer, Cham. https://doi.org/10.1007/978-3-319-19773-9_75.

Hernández, Y., Sucar, L. E., & Conati, C. (2008). *An affective behavior model for intelligent tutors. Proceedings of the 9th international conference on Intelligent Tutoring Systems*, 819–821. Montreal, Canada.

Hsu, C.-L., & Lu, H.-P. (2004). Why do people play on-line games? An extended TAM with social influences and flow experience. *Information and Management*, *41*(7), 853–868.

Hsu, K. C., Lin, K. H. C., Lin, I. L., & Lin, J. W. (2014). The design and evaluation of an affective tutoring system. *Journal of Internet Technology*, 15(4), 533–542.

Jiang, M. (2000). A study of factors influencing students' perceived learning in a web-based course environment. *International Journal of Educational Telecommunications*, *6*(4), 317–338.

Kirsch, D. (1997). The SENTIC MOUSE: Developing a tool for measuring emotional valence. Retrieved August 5 2020, from http://affect.media.mit.edu/projectpages/archived/projects/sentic_mouse.html.

Klein, J., Moon, Y., & Picard, R. W. (2002). This computer responds to user frustration: Theory, design and results. *Interacting with Computers*, *14*(2), 119–140.

Kort, B., Reilly, R., & Picard, R. W. (2001). *An affective model of interplay between emotions and learning: Reengineering educational pedagogy—Building a learning companion. Proceedings of the IEEE International Conference on Advanced Learning Technologies*, 43–48. Madison, USA.

Lin, H. C. K., Chen, N. S., Sun, R. T., & Tsai, I. H. (2014). Usability of affective interfaces for a digital arts tutoring system. *Behavioral and Information Technology*, 33(2), 104–115.

Lin, H. C. K., Su, S. H., Chao, C. J., Hseih, C. Y., & Tsai, S. C. (2016). Construction of a multi-mode affective learning system: Taking affective design as an example. *Educational Technology & Society*, 19(2), 132–147.

Lin, H. C. K., Wang, C. H., Chao, C. J., & Chien, M. K. (2012). Employing textual and facial emotion recognition to design an affective tutoring system. *Turkish Online Journal of Educational Technology – TOJET*, 11(4), 418–426.

Lisetti, C. L. (1999). *A user model of emotion-cognition. Proceedings of the Workshop on Attitude, Personality, and Emotions in User-Adapted Interaction at the International Conference on User-Modeling (UM'99)*. Banff, Canada.

Lowendahl, J.-M. (2017). Hype cycle for education. Retrieved September 1 2020, from http://www.gartner.com/DisplayDocument?doc_cd=233974&ref=g_sitelink

Malone, T. W. (1981). Toward a theory of intrinsically motivating instruction. *Cognitive Science*, *5*(4), 349–361.

Meyer, A. K., Fedor, S., Ghandeharioun, A., Mischoulon, D., Picard, R., & Pedrelli, P. (2020). Feasibility and acceptability of the Empatica E4 sensor to passively assess physiological symptoms of depression. ABCT.

Microsoft Corporation. (2009). Microsoft Agent. Retrieved June 22 2020, from http://www.microsoft.com/products/msagent/main.aspx

Nunnally J.C. (1978). An overview of psychological measurement. In: Wolman B.B. (eds) *Clinical diagnosis of mental disorders.*, Boston, MA: Springer.

Oatley, K. (1992). *Best laid schemes: The psychology of emotions.* Cambridge: Cambridge University Press.

Ortony, A., Clore, G. L., & Collins, A. (1990). *The cognitive structure of emotions.* New York: Cambridge University Press.

Pajares, F. (1996). Self-efficacy beliefs in academic settings. *Review of Educational Research*, *66*(4), 543–578.

Peters, E. (2006). The functions of affect in the construction of preferences. In S. Lichtenstein & P. Slovic (Eds.), *The construction of preference* (pp. 454–463). New York: Cambridge University Press.

Picard, R. W. (1997). *Affective computing.* Massachusetts: MIT Press.

Prendinger, H., Dohi, H., Wang, H., Mayer, S., & Ishizuka, M. (2004). *Empathic embodied interfaces: Addressing users' affective state. Proceedings of the Tutorial and Research Workshop on Affective Dialogue Systems 2004*. Kloster Irsee, Germany, pp. 53–64.

Sarrafzadeh, A., Fan, C., Dadgostar, F., Alexander, S., & Messom, C. (2004). *Frown gives game away: affect sensitive systems for elementary mathematics, Systems, man and cybernetics, 2004 IEEE international conference on*, Vol. 1, pp. 13–18.

Sarrafzadeh, A., Hosseini, H., Fan, C., & Overmyer, S. (2003). *Facial expression analysis for estimating learner's emotional state in intelligent tutoring systems, Advanced Learning Technologies, Proceedings of the 3rd IEEE International Conference on*, pp. 336–337.

Sarrafzadeh, A., Alexander, S., Dadgostar, F., Fan, C., & Bigdeli, A. (2008). "How do you know that I don't understand?" A look at the future of intelligent tutoring systems. *Computers in human behaviour*, 24(4), 1342–1363.

Schwarz, N. (1990). Feelings as information: Informational and motivational functions of affective states. In E. T. Higgins & R. M. Sorrentino (Eds.), *Handbook of Motivation and Cognition: Foundations of Social Behaviour* (pp. 527–561). New York: Guildford Press.

Schwarz, N., & Clore, G. L. (1988). How do I feel about it? The informative function of affective states. In K. Fiedler & I. Forgas (Eds.), *Affect, Cognition, and Social Behavior* (pp. 44–62). Göttingen: Hogrefe.

Shuell, T. J. (1981). Toward a model of learning from instructions. In K. Block (Ed.), *Psychological Theory and Educational Practices: Is it Possible to Bridge the Gap?* Los Angeles, CA: American Educational Research Association.

Sofer, W., & Gribbin, M. (2017). Morgan: A genetics tutorial. Retrieved 1 August, 2020, from http://morgan.rutgers.edu/MorganWebFrames/How_to_use/HTU_Frameset.html.

Sottilare, R. A., & Proctor, M. (2012) Passively classifying student mood and performance within intelligent tutors, *Educational Technology and Society*, 15(2), 101–114.

Stein, N. L., & Levine, L. J. (1991). Making sense out of emotion. In W. Kessen, A. Ortony & F. Kraik (Eds.), *Memories, thoughts, and emotions: Essays in honor of George Mandler* (pp. 295–322). Hillsdale, NJ: Erlbaum.

Steinert, S., & Friedrich, O. (2020). Wired emotions: Ethical issues of affective brain–computer interfaces. *Science and Engineering Ethics*, 26, 351–367. https://doi.org/10.1007/s11948-019-00087-2

Strauss, M., Reynolds, C., Hughes, S., Park, K., McDarby, G., & Picard, R. W. (2005). The HandWave bluetooth skin conductance sensor. In *Affective computing and intelligent interaction*. Berlin/Heidelberg: Springer.

Susarla, S., Adcock, A., Van Eck, R., Moreno, K., & Graesser, A. (2003). *Development and evaluation of a lesson authoring tool for AutoTutor. Proceedings of the Artifical Intelligence in Education Conference*, 378–387. Sydney, Australia.

Thompson, N. (2012). *Development of an open affective computing environment.* Unpublished PhD Thesis, Murdoch University, Perth.

Thompson, N., Koziniec, T., & McGill, T. (2012). *An open affective computing platform. Proceedings of the IEEE 3rd International Conference on Networked and Embedded Systems for Every Application, 1–10.* Liverpool, UK.

Thought Technology. (2010). CardioPro Infiniti HRV analysis module user manual. Retrieved 1 July, 2020, from http://www.thoughttechnology.com/pdf/manuals/SA7590%20CardioPro%20Infiniti %20HRV%20Analysis%20Module%20User%20Manual.pdf

Tian, F., Gao, P., Li, L., Zhang, W., Liang, H., Qian, Y. & Zhao, R. (2014), Recognizing and Regulating E-learners' Emotions Based on Interactive Chinese Texts in E-Learning Systems, *Knowledge-Based Systems*, 55, 148–164.

Walker, E., Rummel, N., & Koedinger, K. (2014), Adaptive Intelligent Support to Improve Peer Tutoring in Algebra, *IJAIED*, 24(1), 33–61.

Woodworth, R. S., & Schlosberg, H. (1954). *Experimental psychology*. New York: Holt.

Yannakakis, G., Hallam, J., & Lund, H. (2008). Entertainment capture through heart rate activity in physical interactive playgrounds. *User modeling and user-adapted interaction*, 18(1), 207–243.

Yuksel, B. F., Oleson, K. B., Harrison, L., Peck, E. M., Afergan, D., Chang, R., & Jacob, R. J. (2016). *Learn Piano with BACh. Proceedings of the 2016 CHI Conference on Human Factors in Computing Systems*. doi:10.1145/2858036.2858388.

3   User State Assessment in Adaptive Intelligent Systems

Jessica Schwarz

CONTENTS

DOI: 10.1201/9781003215349-3

3.1   INTRODUCTION

Adverse mental states of the user can considerably affect the effectiveness and safety of human-machine systems. One example is the aviation accident of Air France Flight AF447 in 2009. The aircraft crashed into the Atlantic Ocean on its route from Rio de Janeiro to Paris killing all 228 passengers and crew members on board. According to the flight accident report (BEA, 2012) the crash was caused by a combination of technical failure and human error. Evidently, the loss of reliable airspeed information due to icing of the pitot probes (airspeed measurement devices) provoked that the autopilot disconnected. These events triggered multiple adverse mental states of the pilots. According to the report, the pilots were *shocked* and *confused* due to the sudden autopilot disconnection. They seemed to be *overloaded* by flying the plane manually in high altitude at night while passing through a thunderstorm and trying to identify the causes for the problem. With erroneous airspeed information and poor visibility, the pilot flying (PF) presumably developed an *incorrect mental model* of the situation. Due to his inadequate "nose up" inputs the aircraft climbed and was destabilized by the resulting loss of air speed. *Distracted* by several permanent alarms and warnings, the pilots, however, did not react to the warning indicating the impending stall.

This airplane accident shows that despite modern, highly automated human-machine systems human and technical failures can occur and, in the worst-case, result in loss of life. The circumstances of the accident suggest that the causes do not only lie in the performance of the technology and the operators but also in the design of the technical system and the human–machine interaction. In the crash of Flight AF447 the adverse user states of the pilots (confusion, overload, incorrect situation awareness, and distraction by alarms) resulted from the interaction with the technical system (sudden disconnection of the autopilot, no support for manual control, insufficient feedback on the causes of the failure, numerous alarms). A different behavior of the technical system may have avoided or reduced these adverse states (cf. Schwarz, Fuchs, & Flemisch, 2014).

Mitigating such critical user states in highly automated systems is the aim of adaptive system design. Adaptive systems assess the current state of the human operator and apply adaptation strategies to counteract critical user states and performance decrements. These approaches have been researched for several decades [e.g., Adaptive Aiding (Rouse, 1988), Adaptive Automation (Scerbo, 1996), and Augmented Cognition (Stanney, Schmorrow, Johnston, Fuchs, Jones et al., 2009) as the most prominent approaches]. More recently, this research area has benefited from technical advances, especially in the field of sensor technology for user state assessment. New possibilities now arise to detect mental states and address them by adaptive technology. Transferring these concepts from the laboratory to practical applications is, however, still challenging. Human-machine systems are impacted by a wide range of influencing factors in the real world that are often not accounted for in laboratory settings (e.g., environmental factors, social context, individual characteristics of the operator). Human error and performance decrements can be the result of multiple interrelated critical user states as seen in the crash of flight AF447. Therefore, some researchers claim that one-dimensional consideration of user state and symptomatic treatment of critical states are not sufficient when designing adaptive systems (e.g., Steinhauser, Pavlas, & Hancock, 2009).

By considering user state as a multidimensional construct, this chapter provides an overview of user state assessment in adaptive system design, by analyzing current research and evaluating available measurement techniques for multiple user state dimensions. It also discusses challenges and implications for the design and implementation of adaptive intelligent systems. Suggesting a holistic view, this chapter introduces a generic model of user state, and it describes the model's integration into RASMUS (Real-time Assessment of Multidimensional User State) – a diagnostic component within a dynamic adaptation framework. Finally, an outlook is provided on how such diagnostics can be employed by an adaptation management component to enable dynamic system adaptation.

3.2 MULTIDIMENSIONAL DEFINITION OF USER STATE

The term "user state" serves as a generic term for mental (cognitive and affective) states of the user in this chapter. Various researchers highlight the importance of considering different qualities of user state when designing human machine systems (Edwards, 2013; Schwarz, Fuchs, & Flemisch, 2014; Steinhauser, Pavlas, & Hancock, 2009). Definitions describing user state as a multidimensional construct are, however, rare. Hockey (2003) and Wilson et al. (2004) established the term operator functional state (OFS) to indicate that user state is a multidimensional phenomenon. In Hockey (2003) OFS is defined as "the multidimensional pattern of processes that mediate performance under stress and high workload, in relation to task goals and their attendant physiological and psychological costs" (p.8). As this definition of OFS focuses on stress and high workload conditions, it does, however, not address other adverse user states, such as fatigue. The "Human Performance Enhancement" approach refers to nine key aspects related to operator performance in the aviation domain including several mental states: mental workload, fatigue, stress, attention, vigilance, situation awareness, team, communication, trust (Edwards, 2013; Silvagni et al., 2015). Likewise, Schwarz, Fuchs, and Flemisch (2014) consider user state as a multidimensional construct by distinguishing six user state dimensions that can considerably affect operator performance: mental workload, engagement or motivation, situation awareness (SA), attention (including vigilance), fatigue, and emotional states (including stress). This chapter refers to this definition when considering user state as it is broader than OFS and not domain specific. Table 3.1 lists, which outcomes of the six state dimensions have shown to impair operator performance and thus should be considered as potentially critical.

TABLE 3.1

Examples of potentially critical outcomes of user state

User state	Potentially critical outcomes	Reference
Mental Workload	Too high, too low	Hancock & Chignell, 1987 Veltman & Jansen, 2006
Engagement	Too low	Deci & Ryan,1985
SA	Incorrect SA	Endsley, 1999
Attention	Incorrect attentional focus/Attentional tunneling	Wickens 2005
	Low vigilance	Mackworth 1948
Fatigue	Too high	May & Baldwin, 2009; Hursh et al., 2004
Emotional State	High arousal, low valence (e.g., stress, anxiety)	Staal, 2004

3.3   STUDIES ON USER STATE ASSESSMENT IN ADAPTIVE SYSTEMS

User state assessment in adaptive system design has been investigated in a variety of domains including aviation, air traffic control, driving, military applications (e.g., Command & Control – C2), process monitoring (e.g., in control centers) as well as training and gaming applications. Table 3.2 shows a selection of research and lists the assessed user states, the application field, and the assessment method(s). Most of this research focuses on mental workload assessment, which is the central element within the Adaptive Automation approach. While research on mental workload assessment addresses numerous operational contexts, research related to other user state dimensions appears to be more domain specific. For example, studies typically assess fatigue during driving tasks, while the assessment of engagement or motivation mainly refers to learning and serious gaming applications. Few studies focus on the simultaneous assessment of different user states.

Table 3.2 also indicates that studies often used a combination of different physiological measures for user state assessment. Haarmann, Boucsein, and Schaefer (2009), for example, combined electrodermal response (EDR) and heart rate (HR) or heart rate variability (HRV) and found that mental workload can be more accurately determined by the combination than by using individual measures. Other studies report a combination of the cardiovascular measures HR and HRV (Mulder, Kruizinga, Stuiver, Vernema, & Hoogeboom, 2004; de Rivecourt, Kuperus, Post, & Mulder, 2008), the oculomotoric measures of blink frequency, fixation frequency and pupil dilation (Van Orden et al., 2001) and a combination of blink, pupil dilation and skin temperature in the face (Wang, Duffy, & Du, 2007). Fuchs et al. (2006) propose a combination of eye movement measures with EEG-based measures to diagnose whether the operator has consciously perceived critical events to evaluate situation awareness. Apart from combining these measures in a rule-based approach, some researchers also employed Artificial Neural Networks or Bayesian Networks to develop classifiers of user state by combining large numbers of different physiological parameters (e.g., Barker & Edwards, 2005, Wilson & Russell, 2003, 2006, and Van Orden et al., 2001). By integrating various parameters using artificial intelligence methods, it was possible to achieve high prediction accuracies. Wilson & Russell (2003)

TABLE 3.2
Overview on assessed user states, application fields, and methods in studies on adaptive system design

User State	Application field	Method	Reference
Mental Workload	Aviation	EEG, EOG	Belyavin, 2005
		EDA, HR, HRV	Haarmann, Boucsein, & Schaefer, 2009
		EEG, ECG	Dorneich et al., 2011
		HRV and cerebral blood flow	Parasuraman, 2003
		HR, HRV, BP, BPV	Veltman & Jansen, 2003
		EEG	Pope et al., 1995
	Air Traffic Control	Performance in secondary task	Kaber et al. 2002; Kaber & Wright, 2003
		EEG, heart rate, eye blinks, respiration	Wilson & Russell, 2003
	Command & Control	HR, HRV, respiration, eye blinks	Veltman & Jansen, 2006
		Various eye tracking measures	De Greef et al., 2009
		Object oriented task model and performance	Arciszewski, de Greef, & van Delft 2009
		Various eye tracking measures	Van Orden et al., 2001
		EEG, ECG, EDA	Tremoulet et al., 2005
	Tracking task	EEG and ERP	Prinzel III, Pope, Freeman et al., 2001
		ERP	Hadley et al., 1999; Scallen et al., 1995
	UAV Operation	Combination of EEG, ECG, EOG, EMG, pupil dilation	Barker & Edwards, 2005
	Driving	Model based assessment	Hancock & Verwey, 1997
	Process monitoring	Cardiovascular measures (e.g., BP, HR, HRV)	Mulder et al., 2004
Situation Awareness	Command & Control	EEG, ERP	Berka et al., 2006; Fuchs et al. 2006
Emotional State	Aviation	HR + facial-EMG	Hudlicka & McNeese, 2002
	Nonspecific	Events, facial expression, theory of mind	Bosse, Memon, & Treur 2008
	Serious Gaming	Facial expression, seating position pressure sensitive mouse, EDA	Woolf et al., 2009
Attention	Command & Control	Eye tracking, Modeling	Bosse et al., 2009
	Text analyses	Eye tracking and EEG	Mathan et al., 2008; Behneman et al., 2009
Fatigue	Driving	EEG	Lin et al., 2006
		Eye blinks; Gaze and head movements, facial expression	Ji, Zhu & Lan, 2004
		Eye blinks and saccades	Schleicher et al., 2008
Motivation	Serious Gaming	Interaction with the system, mouse movements	Ghergulescu & Muntean, 2010, 2011
		Eye Tracking (distraction)	D'Mello et al., 2012
		EEG, EDA	Derbali & Frasson, 2010

(Continued)

TABLE 3.2 (*Continued*)

Overview on assessed user states, application fields, and methods in studies on adaptive system design

User State	Application field	Method	Reference
Mental Workload, Situation awareness	Command & Control	Posture, head position	Balaban et al., 2005; Frank, 2007
Mental Workload, Fatigue	Driving	Steering behavior, blink rate, head position	Hancock & Verwey, 1997
Mental workload, fatigue, attention	Command & Control	HRV, respiration, pupil dilation, mouse click frequency, task-related measures	Schwarz & Fuchs, 2017
Operator Functional State	Driving	43 physiological indicators (e.g., EEG, HR, eye blinks)	Wilson & Russell, 2003
	Operation of UAVs	EEG, EOG, ECG	Wilson & Russell, 2006
	Command & Control	HRV, respiration and others	Gagnon et al., 2014
	Process monitoring	EEG and HRV in combination with performance measures	Ting et al., 2008

Abbreviations: BP: blood pressure; BPV: blood pressure variability; ECG: Electrocardiogram; EDA: electrodermal activity, EEG: Electroencephalography; EMG: electromyogram EOG: Electrooculography; ERP: event-related potentials; HR: heart rate; HRV: heart rate variability.

even report a correct differentiation of 97.5 percent between an overload and non-overload condition.

Hybrid approaches combine different categories of assessment methods. Examples are the combination of performance measures with task-related measures (Hancock & Scallen, 1998, Parasuraman, Mouloua & Molloy, 1996) or the combination with an object oriented task model (Arciszewski, de Greef & van Delft, 2009). Bosse et al. (2009) combined a model-based assessment with eye tracking measures to assess attention. The approach of Schwarz & Fuchs (2017), detailed in Section 3.6, used a combination of physiological, behavioral and task-related measures for the assessment of three adverse user states.

3.4 MEASUREMENT TECHNIQUES

Given that user states are latent constructs that cannot be measured directly, researchers aim at inferring these mental states through various measurement techniques, each being associated with specific strengths and weaknesses. This section focuses on three main categories of empirical measurement techniques: subjective rating, performance-based measures and physiological/behavioral measures.

3.4.1 Criteria to Evaluate Measurement Techniques

To evaluate the strengths and weaknesses of these measurement techniques it is essential to consider which requirements a measurement technique should fulfill when applied in adaptive system design. Evaluations of empirical measurement techniques typically examine the method's psychometric properties. This includes the validity (in this context: ability to measure the user state it is assumed to measure), the reliability/temporal stability (ability to produce consistent and temporarily stable outcomes), the sensitivity (in this context: strength to detect changes of user state) and the diagnosticity (in this context: ability to discriminate between different types of user state). Besides good psychometric properties the method should also provide a continuous assessment of user state and an evaluation in near real time, as the adaptive system needs to act in a timely manner to effectively mitigate critical user states. In applied settings, practicability and freedom from interference are important (Kramer, 1991). As for practicability, it is evaluated how easily the method can be applied and how robust the measurement devices are. It also includes aspects such as cost and training expenditure. Freedom from interference refers to the requirement that the recording should not interfere with the user's task processing and well-being. Ideally, considering multidimensional user state assessment, the method (or combination of methods) must also be able to assess several mental states.

In summary, measurement techniques applied in adaptive system design should ideally meet the following criteria and requirements:

- Good psychometric properties
- Good practicability
- No intrusiveness
- Continuous measurement in real time
- Enable multidimensional assessment

3.4.2 Subjective Rating

Subjective rating techniques typically refer to self-assessments of user state provided by questionnaires, but they can also involve third-party assessments by observing users. Subjective ratings enable a differentiated assessment of various user states. For example, questionnaires are used to assess mental workload (e.g., NASA-TLX, Hart & Staveland, 1988; RSME, Zijlstra, 1993), fatigue (e.g., Stanford Sleepiness Scale, Hoddes, Zarcone, Smythe, Phillips, & Dement, 1973; Karolinska Sleepiness Scale, Åkerstedt, & Gillberg, 1990) or emotional states (e.g., Self-Assessment Manikin (SAM), Bradley, & Lang, 1994). Subjective measures also have high face validity (Cain, 2007). However, their suitability for adaptive systems is limited by the fact that subjective measures do not allow for a continuous assessment without additionally stressing, disturbing or distracting the user during task processing. Ratings may also be biased due to social desirability and response tendencies (Dirican & Göktürk, 2011). Presumably, subjective measures have therefore rarely been used for adaptations (see Section 3.3). However, since subjective rating scales allow for a targeted assessment of different user states and are easy to apply, they can prove useful for validating selected measures.

3.4.3 PERFORMANCE-BASED MEASURES

Performance-based measures include the duration of task completion, the reaction or response time, or the error rate. The assessment can refer either to the primary task or to an embedded or external secondary task (cf. O'Donnell & Eggemeier, 1986). The secondary-task paradigm assesses workload by determining the free capacity of the user that is not involved in completing the primary task. However, secondary tasks, especially if they are not embedded in the operational setting, can pose additional load on the user and might interfere with task processing of the primary task (Grandt, 2004). Performance-based measures offer the advantage that they are objective and easy to apply and to analyze in real time. The measurement itself does not interfere with task processing. However, it depends on the task domain whether performance-based measures also allow for a continuous recording. For example, in driving conditions, it is possible to continuously measure performance, for example, via steering behavior (e.g., Son & Park, 2011; Hurwitz & Wheatley, 2002) while monitoring tasks rarely require actions of the operator indicating his/her current performance. Regarding their application in adaptive systems, it should also be noted, that performance measures do not provide any information about which user states show adverse and potentially critical outcomes.

3.4.4 PHYSIOLOGICAL AND BEHAVIORAL MEASURES

Physiological measures (also called psychophysiological measures in this context) refer to processes in the human body that are influenced by changes in user state, although – as detailed later –other factors than user state can also affect physiological measures (e.g., caffeine, temperature, body movement). Behavioral measures are sometimes equated with performance measures (e.g., Scerbo et al., 2001; Inagaki, 2003). Others (e.g., Elkin-Frankston, Bracken, Irvin, & Jenkins, 2017; Schwarz, 2019) use the term behavioral measures independent from performance for different kinds of motoric activities of the user, such as facial expressions, gestures, posture or operating inputs (e.g., mouse clicks). This section focuses on physiological and (motoric) behavioral measures as they share similar properties.

As the analysis of studies on user state assessment in adaptive system design revealed (cf. Section 3.3) researchers employed various kinds of physiological and behavioral measures. For adaptive system design, physiological and behavioral measures appear particularly relevant, as most of these measures allow for a real-time assessment of user state. They are rather unobtrusive with respect to task completion and can indicate different kinds of user state (Dirican & Göktürk, 2011). This section provides a more detailed analysis on available measures, their relation to different mental states and their strengths and weaknesses.

3.4.4.1 Measures of the Visual System

Parameters of the visual system include eye movements (fixations, saccades), blinking and pupil dilation. Eye tracking is a camera-based measurement technique for recording these parameters. For monitor-based workplaces, remote eye trackers are placed or attached underneath the monitor. They have the advantage to allow contact-free collection of data without interfering with the user. For mobile applications, researchers can use eye trackers integrated in glasses that allow for a rather unobtrusive measurement. Electrooculography (EOG) is another but more intrusive measurement technique, as it requires electrode placement on the scalp similar to EEG.

Measures of the visual system can indicate different user states: attention, fatigue, mental workload, motivation and emotional states. However different states can provoke similar reactions of the visual system (e.g., pupil dilation may increase due to changes in mental workload, emotional states or attention). A differentiation between different user states based on a single measure therefore does not seem practicable.

Furthermore, literature indicates that blink parameters react differently depending on the type of task. Blinks can be suppressed during perceptual tasks in order not to obstruct the reception of visual information. Thus, blink rate was found to decrease with perceptual tasks. In contrast, an increasing frequency of blinks was observed in cognitive tasks and speech motor activities (cf. Hargutt, 2003).

Fixation parameters also show different results depending on whether the task is predominantly visual or cognitive; for example, fixation duration was found to increase on cognitive tasks (Meghanathan, van Leeuwen, and Nikolaev, 2014), but decrease on visual tasks (De Rivecourt et al., 2008). The spatial placement of relevant information also influences eye movements (Grandt, 2004). Thus, when interpreting these measures, there is a need to consider task- and situation-specific factors.

Pupil dilation is less dependent on the type of task, but reacts, among other things, to changes in light (so-called pupil light reflex; Beatty & Lucero-Wagoner, 2000) and caffeine consumption (Stuiber, 2006). Especially light conditions often cannot be kept constant in real-life conditions, such as driving, limiting the applicability of pupil dilation. With the Index of Cognitive Activity (ICA), Marshall (2000, 2002) developed a method that uses short-term and rapid changes in pupil dilation instead of the absolute change in pupil size influenced by light conditions. Marshall (2000, 2002) found that these high-frequency components of pupil dilation react sensitively to changes in mental workload but are independent of light influences. However, the method requires that the pupil dilation is recorded at about 250 Hertz, which is beyond the capabilities of most commercial eye trackers.

3.4.4.2 Measures of Brain Activity

The most prominent technique to retrieve measures of brain activity in studies on user state assessment is electroencephalography (EEG). More recently functional near-infrared spectroscopy (fNIRS) gained popularity for real-time user state assessment. EEG refers to the measurement of electrical brain activity via the scalp. EEG-based measures are sensitive to different mental states including attention, mental workload, emotional states, fatigue, and motivation. Mental states were found to be associated with specific activity patterns in the frequency bands alpha (8–13 Hz), beta (>13 Hz), theta (4–8 Hz) and delta (< 4 Hz). In addition, event-related potentials (ERP), especially the P300 component, can provide information about cognitive responses to certain events that can be used as an indicator for situation awareness (Berka et al., 2006). As EEG-based measures allow for discriminating between different mental states they provide high diagnosticity.

fNIRS is an imaging technique, measuring changes in hemoglobin concentrations within the brain that are associated with cerebral activation. While EEG-based methods have a high temporal but limited spatial resolution, fNIRS has a higher spatial but lower temporal resolution (Strait & Scheutz, 2014). Some researchers have therefore developed combined applications of EEG and fNIRS (e.g., Hirshfield et al., 2009; Nguyen, Ahn, Jang, Jun, & Kim, 2017) to benefit from the advantages of both methods. Although analyses of fNIRS data has mainly been performed offline (Strait & Scheutz, 2014), recent studies evaluated fNIRS for a real-time assessment of mental states (Gateau, Durantin, Lancelot, Scannella, & Dehais, 2015; Hincks, Afergan, & Jacob, 2016).

It should be noted that the measurement of brain activity via EEG and fNIRS is more intrusive than other measurement techniques. It lacks practicability because the equipment is comparably expensive, sensor placement is time consuming, data analysis and interpretation are complex, and it requires special signal processing knowledge.

Considering EEG, "low-cost EEG" have become available in recent years (e.g., Epoc EMOTIV, B-Alert X10, Quasar). They include analysis software that offers automated signal processing as well as algorithms for classifying different mental states in real time. The main advantages are that the headsets are easier to use and no specific background knowledge is required for the analysis and interpretation of the classifiers. However, researchers also report limitations questioning the applicability and the metrics' validity of such low-cost sensors (Noronha, Sol, and Vourvopoulos, 2013; Goldberg, Sottilare, Brawner, and Holden, 2011). Hence, low-cost-sensors and their built-in metrics should be carefully evaluated before employing them for user state assessment within a specific environmental setting.

3.4.4.3 Peripheral-Physiological and Behavioral Measures

Literature indicates that measures controlled by the autonomic nervous system are sensitive to changes in the user state (Dirican & Göktürk, 2011). These measures include cardiovascular measures such as HR, HRV, or blood pressure, as well as measures of skin conductivity, body temperature, and respiration. Most of these measures can be recorded via sensors integrated into breast straps and t-shirts, in-ear sensors or wearables, thus allowing for rather unobtrusive measurements.

Behavioral measures include facial expression, gestures, speech/voice, head movements that can indicate emotional states, motivation, and fatigue. Data acquisition for the analysis of facial expressions can be performed by camera-based methods in a contact-free manner. Another measurement method is electromyography (EMG) of the face, which measures the muscle contractions that cause a facial expression, thus enabling the determination of short-term and barely perceptible changes in facial expression (Mahlke & Minge, 2006). Like EEG, this requires the attachment of electrodes to the skin. Another method is the analysis of speech/voice in which emotional states are reflected (e.g., Lee & Narayanan, 2005; Nwe, Wei & De Silva, 2001). Behavioral measures related to the interaction with the technical system (e.g., mouse clicks or mouse click pressure) can also provide information about the user state. These measures have the advantage that they can be evaluated via log files of the system without the need for an additional sensor (Elkin-Frankston et al., 2017; Hershkovitz & Nachmias, 2011; Ben-Zadok et al., 2009).

Peripheral-physiological and behavioral measures require less effort for measurement and evaluation compared to measures of brain activity. However, peripheral-physiological measures appear to have less diagnosticity as they mostly reflect arousal, which can be influenced by different user states (Hogervorst, Brouwer & van Erp, 2014).

The influence of other state-independent factors on peripheral-physiological activity must also be taken into account. For example, the heart rate reacts to physical activity (Mulder, 1992), caffeine consumption (Barry et al., 2005), or changes in climatic conditions (Grandt, 2004). HRV is influenced by body position, circadian rhythm, diet, age, and in particular, respiration (Uhlig, 2018). The influence of respiration on HRV is also known as respiratory sinus arrhythmia (breath-synchronous variation of heart rate). It is therefore recommendable to measure both, respiratory rate and depth, in order to interpret changes in HRV correctly (Mulder, 1992). Since speech affects breathing, this also has an impact on HRV and should be considered when interpreting these measures (Veltman & Gaillard, 1998).

3.4.5 CONCLUSION

Focusing on a multidimensional and domain-independent consideration of user state assessment, this section provided a concise overview on measurements techniques. More detailed information on the assessment of specific user states is provided by reviews on: *Fatigue*: Barr, Popkin & Howarth (2009), Kecklund et al. (2007), Wright, Stone, Horberry, & Reed (2007), and Lal & Craig (2001); *Mental workload*: Kramer (1991), Miller (2001), Cain (2007), and Farmer & Brownson (2003); *Attention/vigilance*: Oken, Salinsky, & Elsas (2006); *Emotional states*: Mauss & Robinson (2009) and Calvo & D`Mello (2010); and *Motivation*: Touré-Tillery & Fishbach (2014). Table 3.3 summarizes the main advantages and disadvantages of each group of measurement technique for use in adaptive systems.

Overall, subjective measures appear inappropriate for a real-time diagnosis of user state, as regular measurement would disturb the operator during task processing. Performance measures are less intrusive, but considering a multidimensional user state assessment, performance-based measures do not provide any information on the potential causes for performance decrements (adverse user states and impact factors). In comparison, physiological and behavioral measures appear to be more appropriate for a real-time assessment of user state in adaptive systems. Physiological and behavioral measures mostly allow for a continuous data acquisition and assessment of the data in real time. It is also possible to assess different user states. In addition, sensors (e.g., eye trackers, webcams, wearables) are becoming less intrusive, thus, providing recordings with little impairment of the operator during task processing (Dirican & Göktürk, 2011). It is therefore not surprising that studies on user state assessment in adaptive system design often employ physiological and behavioral measures (cf. Table 3.2).

As outlined earlier, there are also challenges when using physiological and behavioral measures. Most physiological and behavioral measures are also influenced by factors other than user state, so that these factors must either be controlled or included in the diagnosis. Literature also indicates that the sensitivity and temporal stability

TABLE 3.3

Advantages and disadvantages of empirical measurement techniques for user state assessment in adaptive systems

Method	Advantages	Disadvantages
Subjective Rating Techniques	• Easy to apply • Cheap (mostly freely available) • Established questionnaires mostly have good psychometric properties	• May induce additional load and distraction • May be biased due to social desirability • No continuous measurement without distracting the user
Performance based measures	• Easy to apply • Cheap • No interference of the operator • Continuous assessment (task-dependent)	• No diagnosticity → no discrimination between different user states • No continuous assessment for monitoring tasks
Physiological and behavioral measures	• Continuous measurement • Real-time assessment • Sensors are becoming less intrusive as technology advances • Potentially high diagnosticity if combined with other measures	• Costs for sensors can be high • Intrusiveness can be high depending on the sensor • Impact factors other than user state • Sensitivity, Reliability, and temporal stability may vary between different situations and individuals

varies between and within individuals (cf. Schwarz & Fuchs, 2016). Section 3.5 addresses these challenges and implications for adaptive system design.

3.5   IMPLICATIONS FOR ADAPTIVE SYSTEM DESIGN

This section summarizes major challenges and provides recommendations for adaptive system design. It also introduces a holistic view on user state as an approach to face these challenges.

3.5.1   CHALLENGES AND RECOMMENDATIONS

3.5.1.1   Confounding Factors

The evaluation of measurement techniques (see Section 3.4) showed that different user states could provoke similar outcomes in the physiological reactions. This makes it difficult to discriminate between different user states. But also other factors not directly related to the user state influence physiological reactions. These include the type of task, lighting conditions, caffeine consumption, physical activity, or speech (see Section 3.4.4). In terms of user state assessment, these factors represent confounding influences, as they can mask changes caused by user state (Kramer, 1991). In laboratory settings, it may be possible to eliminate these interfering factors or keep them constant. However, this is not always possible in real-world settings. Brouwer, Zander, van Erp, Korteling, and Bronkhorst (2015) recommend assessing known confounding factors and to ensure that they do not lead to incorrect diagnostic decisions.

3.5.1.2 Reliability and Temporal Stability

Physiological and behavioral measures can be influenced by a variety of known and unknown factors. In addition to factors which systematically influence the measurement result, artifacts, varying personal stress factors, or changes over time can limit reliability and stability (Byrne & Parasuraman, 1996). The findings of Faulstich et al. (1986) and Tomarken (1995) indicate that temporal stability varies between diagnostic measures. For application in adaptive system design, it is therefore advisable to assess not only the validity but also the temporal stability of physiological measures and to perform these analyses at an individual level (Schwarz & Fuchs, 2016; see also next subsection on individual differences).

Many studies in the context of adaptive system design have used a combination of different measures or a combination of different methods in hybrid models to obtain a more robust diagnosis (cf. Section 3.3). Combined measures offer the possibility that unreliable assessments of an individual measure can be compensated by other measures so that the robustness and reliability of the diagnosis can be significantly increased (Stanney et al., 2009). This has also been demonstrated experimentally (e.g., Wilson & Russell, 2003; Haarmann, Boucsein, & Schaefer, 2009).

3.5.1.3 Individual Differences

When using physiological and behavioral measures, it must be considered that individuals may also differ in their physiological and behavioral responses regardless of their mental state. Examples are anatomical differences in pupil size, differences in (resting) pulse or blink frequency. It is therefore recommendable to compare physiological measures to an individual baseline (measured values of an individual at rest) (Scerbo, 2001).

Furthermore, it is necessary for the technical system to be able to diagnose differences in the mental state of an individual operator in real time (so-called "single-trial" diagnosis) to be able to react to adverse states in a timely manner (cf. Section 3.4.1). Evaluations at group level, as they are usually carried out in experimental studies, are not meaningful in this context. On one hand, the measures' sensitivity to changes in the user's state can vary between individuals (Veltman & Jansen, 2003). On the other hand, individuals cope with task demands differently, resulting in interindividual differences in user state and performance (see Belyavin 2005, Parasuraman et al., 1992; Veltman & Jansen, 2003). It is therefore possible that a parameter may indicate statistically significant changes in user state at the group level but not react sensitively at the individual level (cf. Hogervorst, Brouwer, & van Erp, 2014; Schwarz & Fuchs, 2016). The suitability of diagnostic measures must therefore be evaluated at the individual level.

3.5.1.4 Self-Regulation Strategies of the Human Operator

As Hancock and Chignell (1987) and Veltman and Jansen (2004, 2006) point out, the human is also an adaptive system that continuously adapts to changing task requirements. Physiological reactions that indicate increased mental effort can be a sign of a successful adaptation of the operator to changing task demands. It does not necessarily mean that he/she needs support. On the contrary, it might reflect a high involvement of the operator in the task. If these changes in workload were used by the technical system to reduce the task load, it might result in counterproductive interaction between the two adaptive systems (Veltman & Jansen, 2006). However, as accidents such as the crash of flight AF447 indicate, there are also situations when the operator's state regulation processes fail to successfully maintain the operator's effectiveness. Veltman and Jansen (2004, 2006) propose that adaptive technical systems are more likely to work successfully if adaptation is triggered only in those situations when the operator is no longer able to adapt adequately to changing task demands. An indication could be that he/she ignores certain tasks or parts of tasks.

3.5.1.5   Root-Cause Analysis

Indicators of user state such as physiological and behavioral measures usually reflect reactions of the user to changing task conditions. They do not provide any information on the causes for adverse user states. As pointed out by Sciarini and Nicholson (2009) and Steinhauser et al. (2009) a symptom-based adaptation is not sufficient to mitigate adverse user states effectively. User state assessment should therefore also analyze possible causes for critical user states. Various researchers claim that adaptive system design needs to consider context and environmental factors as well as individual characteristics of the operator in addition to the user state (Whitlow & Hayes, 2012; Steinhauser et al., 2009; Fuchs et al., 2006). According to Fuchs et al. (2006), context information and task status are particularly relevant to decide not only when to adapt but also what and how to adapt. Feigh et al. (2012) list a classification of different factors that can be used as triggers for adaptation in addition to user state. A distinction is made between system-based, environment-based, task- and mission-based and spatio-temporal triggers.

3.5.1.6   Oscillation

Physiological measures often have a high sensitivity to change, which means that the recorded values oscillate and may fall below and above the threshold value that triggers adaptation in short time intervals (Barker, Edwards, O'Neill, & Tollar, 2004; Inagaki, 2003). It has been shown that this oscillation, called "yo-yoing" (Stanney et al., 2009; Diethe, 2005), has undesirable effects on the performance of the operator, as it can lead to confusion and might increase mental workload (Stanney et al., 2009). To counteract oscillation, it is proposed to smooth the physiological measurements, by applying filters (Diethe et al., 2004) or to average the data over time intervals (Roscoe, 1993; Haarmann et al., 2009).

It should also be considered that an oscillation of the adaptation in a control loop system can be caused by the adaptation itself. For example, if the mental workload is critically high, the adaptation begins. Subsequently, the adaptation reduces the workload level until it falls below a threshold value and the adaptation is suspended. This can lead to a situation where workload increases again and the adaptation must be triggered again (cf. Stanney et al., 2009). To avoid this problem, Barker et al. (2004) propose to suspend adaptation only after a task that has led to high workload has been completed. The system therefore has to recognize which task the operator is currently working on and when it is completed.

3.5.1.7   Realistic Work Environments

Experimental studies often use artificial tasks to evoke and investigate specific mental states (e.g., oddball task, n-back task). The transferability of these findings to operative conditions is limited, however, as they do not reflect the dynamics and the task complexity of the real world and are often of comparatively short duration. Mulder et al. (2008) point out that (semi-)realistic working conditions should be established for the investigation of state changes that only occur after a longer period, as is the case with fatigue. Near realistic working conditions can be created by simulators. Driving simulators are used in the automotive sector and cockpit simulators in aviation, in which the usual operational tasks can be performed realistically and without risk to the environment.

3.5.1.8 Summary

Table 3.4 summarizes the challenges and recommendations detailed above considering user state assessment in adaptive system design.

3.5.2 HOLISTIC VIEW ON USER STATE

Schwarz, Fuchs, and Flemisch (2014) suggest that applying adaptive systems to real-world settings requires a holistic view on user state. The authors present a generic model of user state aiming at providing such a holistic view. Figure 3.1 provides a simplified version of the model. The model refers to the findings detailed in the previous section and includes findings from a literature analysis that focused on psychological theories and models of user state (cf. Schwarz, 2019). The model's components and the interrelations between these components are briefly described below.

- *Current user state* refers to the multidimensional user state with the six state dimensions, introduced in Section 3.2. As indicated by the arrows between these user states, empirical findings and psychological theories suggest

TABLE 3.4
Challenges and design recommendations

Nr.	Challenging aspect	Recommendation
1	Confounding factors	• Assess factors that affect the assessment
2	Reliability and Temporal Stability	• Evaluate the temporal stability of physiological indicators • Combine different physiological and behavioral measures to increase the robustness
3	Interindividual differences	• Perform analyses at an individual level • Compare to an individual baseline
4	Self-Regulation Strategies	• Only adapt, when the human's self-regulation failed (or is about to fail)
5	Root-Cause Analysis	• Identify causes for adverse user states by analyzing relevant impact factors
6	Oscillation	• Define a minimum time interval between adaptation changes • Apply methods to smoothen physiological measures
7	Realistic setting	• The experimental setting should provide realistic tasks and a realistic task environment (e.g., by using simulators).

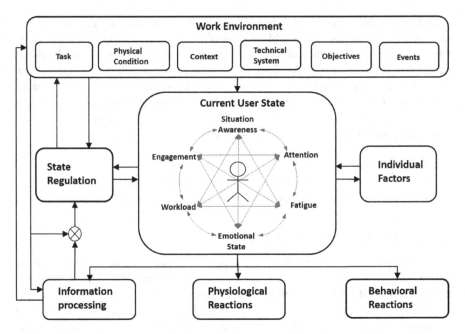

FIGURE 3.1   Generic model of user state (adopted from Schwarz, Fuchs & Flemisch 2014).

interrelations between state dimensions (cf. Edwards, 2013 and Schwarz, 2019). For example, attention can not only be diminished by fatigue but also narrowed by strong emotional states such as anxiety, or by high workload and stress.

- *Work environment* lists external factors that can have an impact on user state. The model distinguishes between task-related factors (e.g., number, complexity, urgency of tasks), factors of the physical environment (e.g., temperature, noise, lighting conditions), context factors (e.g., time of day or social context), properties of the technical system (e.g., level of automation, usability), objectives of the human-machine system, and events.
- *Individual factors* refer to user-specific factors that can affect user state and performance as well. They can be rather long-term factors such as capabilities, experience, knowledge or short-term factors such as well-being, personal needs, or the quality and amount of sleep.
- User state affects *Physiological and Behavioral Reactions* (cf. Section 3.4.4) as well as *Information Processing*. Considering the processing of cognitive tasks, outputs of the information processing component result in task performance. Hence, user states that impair the quality and intensity of information processing, such as stress or fatigue, are likely to impair task performance.
- *State Regulation* refers to the finding of Section 3.5.1.4 that the user can apply self-regulation strategies to maintain his/her level of performance. The gateway (X) in the model indicates that the human compares the target performance required for meeting the objectives with the actual performance. State regulation aims at reducing any discrepancies, for example, by adjusting the

level of effort (cf. Veltman & Jansen, 2006). However, as pointed out in Schwarz, Fuchs, and Flemisch (2014), state regulation may also address other adverse user states. For example, if a car driver recognizes that he/she is tired, he/she may decide to drink more coffee or take a break.

This generic model of user state can serve as a basic framework for a holistic consideration of user state in adaptive system design. As this is a generic model, applying the model to a specific task domain still requires the identification of relevant impact factors and user states. An exemplary approach illustrating how the model's components can be addressed in a real-time assessment of user state and how they can support dynamic system adaptation is presented with RASMUS in Section 3.6.

3.6   REAL-TIME ASSESSMENT OF MULTIDIMENSIONAL USER STATE (RASMUS)

RASMUS is the diagnostic component within a dynamic adaptation framework. Figure 3.2 shows a simplified model of the adaptation framework (adopted from Schwarz & Fuchs, 2017). The adaptive technical system comprises two main components: *information processing* and *state regulation*. The information processing component analyses and displays data from the environment and processes operator inputs, thus representing the basic functionality of traditional technical systems in human–machine interaction. The state regulation component enables adaptive behavior of the technical system. Similarly to the human information processing detailed in Parasuraman, Sheridan, and Wickens, (2000), this component includes four stages of state regulation: (1) data acquisition, (2) user state assessment, (3) action selection, and (4) execution. RASMUS diagnostics address the first two stages of state regulation: the acquisition of data from the operator and the environment and the subsequent assessment of the user state. The stages of action selection and execution refer to the selection and application of appropriate adaptation strategies, accomplished through an Advanced Dynamic Adaptation Management (ADAM, cf. Fuchs & Schwarz, 2017).

3.6.1   DIAGNOSTIC CONCEPT

User state assessment in RASMUS focuses on two main objectives: (1) determine when the user needs support (detect an adverse user state that triggers adaptation) and (2) provide diagnostics for identifying potential causes for adverse states. The adaptation management requires this information to determine what kind of support is most appropriate to restore the user's performance (see Section 3.6.3). For accomplishing these objectives, the diagnostic process of RASMUS consists of four consecutive steps, presented in Figure 3.3 and described below (for a more detailed description of the diagnostic process see Schwarz & Fuchs, 2017).

Step1: RASMUS acquires data from the user, the technical system and the environment that are necessary for the subsequent analyses in steps 2 and 3. RASMUS merges and synchronizes these data streams in real time and normalizes the physiological data on individual baseline states.

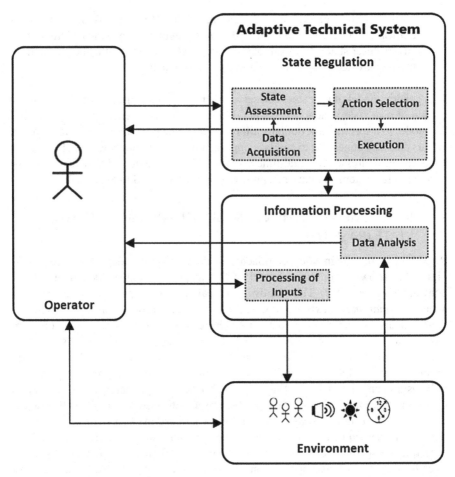

FIGURE 3.2 Model of the dynamic adaptation framework (adopted from Schwarz & Fuchs 2017.)

Step 2: Unlike other approaches of adaptive system design, RASMUS derives a need for support not from physiological data but from performance measures. This means the adaptive system triggers an adaption only at those times, when RASMUS detects a so-called performance decrement. While changes in physiological reactions may be due to productive self-regulation strategies of the user, a performance decrement is a clear indication that the user's self-regulation has failed, and the user needs support. Thus, it ensures that adaptation of the technical system does not interfere with the user's productive self-regulation strategies.

Step 3: If RASMUS detects a performance decrement, it then analyzes potential causes for the performance decrement. Literature and empirical findings suggest combining different kinds of measures in order to gain more robust and valid diagnoses (cf. Section 3.5.1.2). RASMUS therefore analyzes outcomes of physiological and behavioral data, data from the work environment, and individual factors to identify adverse user states that may have provoked the performance decrement.

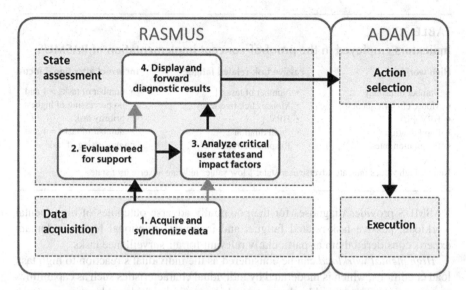

FIGURE 3.3 Diagnostic process in RASMUS (adopted from Schwarz & Fuchs 2017, 2018) (gray arrows indicate data flow between steps.)

Step 4: RASMUS forwards all diagnostic results to the adaptation management component where they are processed to select, configure, and execute appropriate adaptation strategies (cf. Section 3.6.3).

3.6.2 IMPLEMENTATION FOR AN ANTI-AIR-WARFARE TASK

Schwarz and Fuchs (2017 and 2018) applied and tailored the generic diagnostic framework of RASMUS to the specific requirements of a naval Anti-Air-Warfare (AAW) task paradigm. For this purpose, it was necessary to determine appropriate indicators for performance and user state assessments and to specify the rule base for critical outcomes.

3.6.2.1 Task Environment

RASMUS diagnostics were implemented as a Java-based research test bed and connected to an existing AAW simulation. The simulation includes four simplified AAW tasks with different priorities. These tasks occur at scripted times throughout the scenario and may also occur simultaneously. In this case the task with the highest priority must be performed first. Each task is associated with a time limit for task completion. If a task is not completed within the time limit or if task completion is incorrect, RASMUS detects a performance decrement (cf. Schwarz & Fuchs, 2017 for a more detailed description).

3.6.2.2 User States and Indicators

The proof of concept implementation focused on three out of the six user state dimensions introduced in Section 3.2: workload, attention, and fatigue. Specifically,

TABLE 3.5

Indicators employed in the proof-of-concept implementation of RASMUS

High workload	Passive task-related fatigue	Incorrect attentional focus
• Number of tasks ↑ • Mouse click frequency ↑ • HRV ↓ • Pupil dilation ↑ • Respiration rate ↑	• Number of tasks ↓ • Mouse click frequency ↓ • HRV ↑ • Pupil dilation ↓ • Respiration rate ↓	• number of tasks > 1 and no processing of highest priority task • number of tasks = 1 and no processing of task

Note: ↑ high values indicate adverse user state; ↓ low values indicate adverse user state.

RASMUS provides diagnoses for the potentially adverse outcomes of high mental workload, passive task-related fatigue, and incorrect attentional focus as domain experts considered them as particularly relevant for air surveillance tasks.

High mental workload can be considered as the individual's reaction to high task load conditions, which is moderated by individual characteristics such as capabilities or experience. Thus, workload may vary between individuals under the same task load conditions. For diagnosing high mental workload on an individual level, RASMUS uses a combination of five physiological, behavioral, and task-related indicators listed in Table 3.5.

Passive task-related fatigue results from processing of monotonous tasks or monitoring tasks with low cognitive demand over a longer period (cf. May & Baldwin, 2009). It is associated with a low level of arousal resembling the contrary adverse user state to high workload. The assessment of passive task-related fatigue in RASMUS is based on the same indicators as of high workload but with opposite criteria (see Table 3.5).

The operationalization of *incorrect attentional focus* is based on Wickens' concept of "attentional tunneling". Wickens (2005) defines attentional tunneling as the "allocation of attention to a particular channel of information, diagnostic hypothesis or task goal, for a duration that is longer than optimal, given the expected cost of neglecting events on other channels, failing to consider other hypotheses, or failing to perform other tasks" (Wickens, 2005, p. 1). RASMUS detects incorrect attentional focus if a higher priority task is neglected because the user is processing a lower priority task or if he/she missed a task in the absence of an alternative task.

3.6.2.3   Rule Base

For high workload and passive task-related fatigue, RASMUS labels a state as critical if at least 3 out of 5 indicators show critical outcomes. Critical outcomes of each indicator are evaluated for moving mean windows of 30 seconds in order to smooth the data (cf. the challenge of "oscillation" in Section 3.5.1.6). Number of tasks and number of mouse clicks, values are considered critical if they fall above predetermined thresholds that were determined based on previous experimental results (see Schwarz & Fuchs 2017 for a more detailed description).

To account for individual differences in physiological reactions, RASMUS detects critical outcomes of physiological measures by analyzing the deviation of current recordings from a baseline. The physiological indicator is labeled as critically high or low if the current mean deviates by more than 1 SD from the baseline mean.

3.6.2.4 Validation

A validation experiment examined the validity of RASMUS diagnostics provided by this proof-of-concept implementation for the detection of high workload, passive task-related fatigue, and incorrect attentional focus. For validation purposes, the diagnostic outcomes of RASMUS were compared with subjective comparative measures at the time of performance decrements (see Schwarz & Fuchs, 2018 and Schwarz, 2019 for a detailed description). Although facing some limitations, such as a small sample size, results confirm the congruent validity of RASMUS diagnostics for states of high workload, passive task-related fatigue, and incorrect attentional focus. The temporal stability of these outcomes could be confirmed by a later follow-up experiment (cf. Bruder & Schwarz, 2019) that was focusing on evaluating and optimizing the diagnostic rules employed by RASMUS for the classification of critical indicators and adverse user states.

3.6.3 Using RASMUS Diagnostics for Adaptation Management

Adaptation Management is essential in order to avoid cognitive costs associated with adaptation strategies such as task switching issues, situation awareness problems, and increases in workload. These costs may outweigh their benefits and impair performance if adaptations are not designed or employed properly (Fuchs et al., 2007). To address this challenge, Fuchs & Schwarz (2017) designed the Adaptation Management component ADAM that dynamically selects, configures, and applies context-sensitive adaptation strategies.

As illustrated in Figure 3.4, ADAM utilizes RASMUS' diagnostic outcomes on user states and contextual factors (outcomes of specific user state indicators) to determine an adaptation objective (1) and then to select (2) and to configure (3) a suitable adaptation strategy to address the adaptation objective given the current situational context. Subsequently, ADAM triggers the adaptation strategy (4) and monitors the impact of adaptation (5) to determine whether and how adaptation should be continued.

Conceptually, ADAM supports multiple adaptation objectives per critical user state and multiple strategies per adaptation objective in order to better accommodate the implications of varying contextual factors. For example, a diagnosed high-workload state may refer to a response bottleneck caused by high task load (e.g., high number of simultaneous tasks). However, high workload could also be caused by problems with decision-making (executive function bottleneck). ADAM uses context information (e.g., number of tasks, number of mouse clicks) provided by RASMUS to identify the potential cause for the high workload state and to select appropriate adaptation objectives (e.g., reduce task load vs. support decision-making). ADAM then determines the best candidate from a pool of adaptation strategies for a given situation. For example, reducing task load may be accomplished by an Automation strategy that offloads low-priority tasks to the technical system to free operator cognitive resources for high priority tasks. Alternatively, Context-Sensitive Help, indicating the next step of action, could be an effective adaptation strategy to support the user in his/her decision-making and reduce the executive function bottleneck.

Five adaptation strategies were implemented and evaluated for the AAW task introduced in Section 3.6.2: Context-Sensitive Help, Automation, Scheduling, Visual

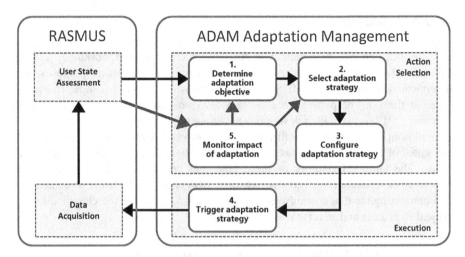

FIGURE 3.4   Five steps of dynamic adaptation management (adopted from Fuchs & Schwarz 2017).

Cueing, and Decluttering (cf. Fuchs et al., 2019, Fuchs et al., 2020). These strategies are designed to address cognitive bottlenecks resulting from high workload and incorrect attentional focus. One experiment aimed at evaluating the effectiveness of each adaptation strategy individually. While two strategies significantly reduced the average duration of critical user state episodes and two others showed promising statistical trends in the same direction, the overall results suggest that the individual effectiveness of the adaptations may be limited by their highly specific respective purposes. A follow-up experiment that employed all five adaptation strategies in dynamic interplay showed that the same strategies indeed complemented each other effectively within a Dynamic Adaptation Management approach (cf. Fuchs et al., 2019, Fuchs et al., 2020).

3.7   CONCLUDING REMARKS

Adapting technical systems to the user's mental state is a promising approach to mitigate performance breakdowns of human-machine systems. However, real-world conditions are a complex interplay of different external conditions, task demands, actions of the technical system, and actions of the human operator. As the crash of Air France flight AF447 has illustrated, these factors can provoke different kinds of adverse user states affecting operator performance. Consideration of these impact factors and adverse user states is critical for designing adaptive intelligent systems that effectively support operator performance. Following a multidimensional consideration of user state, focusing on six user state dimensions (mental workload, engagement, situation awareness, attention, fatigue, and emotional states), this chapter provided an overview on the state of the art, available measurement techniques, and challenges related to the (multidimensional) user state assessment in adaptive system design.

It was shown that there is no single best method available to provide real-time assessments of user state for adaptive systems, but that it is advisable to combine different kinds of measures. Combined assessments can help increasing the robustness of the assessment and enable discrimination between different user states. Moreover, the integration of impact factors and context information into the assessment is necessary in order to be able to tailor adaptation strategies to the specific situation and address causes rather than symptoms of performance problems. A holistic view on user state is therefore essential to design adaptive systems that provide adequate support in a given situation and enable the technical system to interact with the human operator more like cooperative social actors than technical artifacts.

Following this holistic view, this chapter introduced RASMUS and ADAM as recent examples of a diagnostic and an adaptation management component that aim at addressing the identified challenges in adaptive system design through dynamic context-sensitive adaptation. A proof-of-concept implementation demonstrated the applicability and usefulness of these approaches for a realistic air surveillance task. The implementation was limited by focusing on a small number of adverse user states and impact factors. A complete assessment of all relevant user states and impact factors in real-world settings is challenging but does not seem impossible in the future in view of the presumed advances in sensor technology allowing for more robust, artifact-free, and unobtrusive assessments as well as advances in artificial intelligence leveraging data aggregation and interpretation of big data sets in real time. This implies, however, that, as in all everyday applications, ethical questions concerning the collection and interpretation of large amounts of user-specific data also must be addressed.

REFERENCES

Åkerstedt, T., and Gillberg, M. 1990. Subjective and Objective Sleepiness In The Active Individual. *International Journal of Neuroscience* 52(1–2): 29–37. doi:10.3109/00207459008994241.

Arciszewski, H. F. R., T. E. de Greef, and J. H. van Delft. 2009. Adaptive Automation in A Naval Combat Management System. *IEEE Transactions on Systems, Man, and Cybernetics - Part A: Systems and Humans* 39(6): 1188–1199. doi:10.1109/tsmca.2009.2026428.

Balaban, C. D., J. Prinkey, G. Frank, and M. Redfern. 2005. Postural measurements seated subjects as gauges of cognitive engagement. In *Proceedings of the 11th International Conference on Human-Computer Interaction (Augmented Cognition International)*, ed. D.D. Schmorrow, 321–328. New York: Lawrence Erlbaum.

Barker, R. A., and R. E. Edwards. 2005. *The Boeing Team Fundamentals of Augmented Cognition. Proceedings of the 1st International Conference on Augmented Cognition*, 469–476.

Barker, R. A., Edwards, R. E., O'Neill, K. R., and Tollar, J. R. 2004. *DARPA Improving Warfighter Information Intake Under Stress - Augmented Cognition Concept Validation Experiment (CVE). Analysis Report for the Boeing Team*. Arlington: DARPA.

Barr, L., S. Popkin, and H. Howarth. 2009. *An Evaluation of Emerging Driver Fatigue Detection Measures and Technologies*. Final Report FMCSA-RRR09-005. Federal Motor Carrier Safety Administration, Washington DC, USA.

Barry, R., J. Rushby, M. Wallace, A. Clarke, S. Johnstone, and I. Zlojutro. 2005. Caffeine effects on resting-state arousal. *Clinical Neurophysiology* 116, 11: 2693–2700. doi:10.1016/j.clinph.2005.08.008.

BEA. 2012. *Final Report on the accident on 1st June 2009 to the Airbus A330-203 registered F-GZCP operated by Air France flight AF447 Rio de Janeiro – Paris*. Paris: Bureau d'Enquêtes et d'Analyses pour la sécurité de l'aviation civile. Retrieved August 16, 2020 from: http://www.bea.aero/docspa/2009/f-cp090601.en/pdf/f-cp090601.en.pdf.

Beatty, J., and B. Lucero-Wagoner. 2000. The pupillary system. In *Handbook of psychophysiology*, eds. J. T. Cacioppo, L. G. Tassinary, and G. G. Berntson, pp. 142–162. New York: Cambridge University Press.

Behneman, A., N. Kintz, R. Johnson, C. Berka, K. Hale, S. Fuchs, P. Axelsson, and A. Baskin. 2009. Enhancing Text-Based Analysis Using Neurophysiological Measures. *Foundations of Augmented Cognition. Neuroergonomics and Operational Neuroscience*, 449–458. doi:10.1007/978-3-642-02812-0_53.

Belyavin, A. 2005. Construction of appropriate gauges for the control of Augmented Cognition systems. In *Proceedings of the 1st International Conference on Augmented Cognition*, Las Vegas, NV, 22–27.

Ben-Zadok, G., A. Hershkovitz, E. Mintz, and R. Nachmias. 2009. Examining online learning processes based on log files analysis: A case study. *Proceedings of the Fifth International Conference on Multimedia and ICT in Education*, *1*: 55–59.

Berka, C., D. J. Levendowski, G. Davis, M. Whitmoyer, K. Hale, and S. Fuchs. 2006. Objective measures of situational awareness using neurophysiology technology. In *Foundations of Augmented Cognition*, eds. D. D. Schmorrow, K. M. Stanney, and L. M. Reeves, 145–154. Arlington, VA: Strategic Analysis, Inc.

Bosse, T., Z. A. Memon, and J. Treur. 2008. Adaptive Estimation of Emotion Generation for an Ambient Agent Model. In *AmI 2008, LNCS, vol. 5355*. eds. E. Aarts, J. L. Crowley, B. de Ruyter, H. Gerhauser, A. Pflaum, J. Schmidt et al., 141–156. Heidelberg: Springer.

Bosse, T., R. van Lambalgen, P. P. van Maanen, and J. Treur. 2009. Attention Manipulation for Naval Tactical Picture Compilation. In *Proceedings of the 9th IEEE/WIC/ACM International Conference on Intelligent Agent Technology, IAT'09 (Vol. 2)*, eds. R. Baeza-Yates, J. Lang, S. Mitra, S. Parsons, and G. Pasi, 450–457. Washington, DC: IEEE Computer Society Press.

Bradley, M. M., and P. J. Lang. 1994. Measuring emotion: The self-assessment manikin and the semantic differential. *Journal of Behavior Therapy and Experimental Psychiatry* *25(1)*: 49–59.

Brouwer, A. M., T. O. Zander, J. B. van Erp, J. E. Korteling, and A. W. Bronkhorst. 2015. Using neurophysiological signals that reflect cognitive or affective state: six recommendations to avoid common pitfalls. *Frontiers in Neuroscience*, *9*: 136. doi: 10.3389/fnins.2015.00136.

Bruder, A., and J. Schwarz. 2019. Evaluation of Diagnostic Rules for Real-Time Assessment of Mental Workload within a Dynamic Adaptation Framework. In *Adaptive Instructional Systems. HCII 2019. Lecture Notes in Computer Science, vol 11597*, eds. R. Sottilare, and J. Schwarz, 391–404. Cham: Springer.

Byrne, E., and R. Parasuraman. 1996. Psychophysiology and adaptive automation. *Biological psychology*, *42(3)*: 249–268.

Cain, B. 2007. *A review of the mental workload literature*. Toronto: Defence Research and Development.

Calvo, R. A., and S. D'Mello. 2010. Affect detection: An interdisciplinary review of models, methods, and their applications. *IEEE Transactions on affective computing*, *1(1)*: 18–37.

D'Mello, S., A. Olney, C. Williams, and P. Hays. 2012. Gaze tutor: A gaze-reactive intelligent tutoring system. *Journal of Human Computer Studies*, *70(5)*: 377–398. doi:10.1016/j.ijhcs.2012.01.004

De Greef, T., H. Lafeber, H. Van Oostendorp, and J. Lindenberg. 2009. Eye Movement as Indicators of Mental Workload to Trigger Adaptive Automation. In *Foundations of Augmented Cognition. Neuroergonomics and Operational Neuroscience*, eds. D. Schmorrow, I.V. Estabrooke, and M. Grootjen, 219–228. Heidelberg: Springer.

De Rivecourt, M., M. N. Kuperus, W. J. Post, and L. J. M. Mulder. 2008. Heart rate and eye movement measures as indices for mental effort during simulated flight. *Ergonomics*, *51(9)*: 1295–1319.

Deci, E. L., and R. M. Ryan. 1985. *Intrinsic motivation and self-determination in human behavior*. New York: Plenum.

Derbali, L., and C. Frasson. 2010. *Prediction of Players Motivational States Using Electrophysiological Measures during Serious Game Play. Conference on Advanced Learning Technologies, IEEE International*, Sousse, Tunisia: 498–502.

Diethe, T. 2005. The future of augmentation managers. In *Foundations of augmented cognition*, ed. D. D. Schmorrow, 631–640. Mahwah, NJ: Erlbaum.

Diethe, T., B. T. Dickson, D. Schmorrow, and C. Raley. 2004. Toward an augmented cockpit. In *Human performance, situation awareness and automation: Current research and trends* (Vol. 2), eds. D. A. Vicenzi, M. Mouloua, and P. A. Hancock, 65–69. Mahwah, NJ: Erlbaum.

Dirican, C., and M. Göktürk. 2011. Psychophysiological measures of human cognitive states applied in human computer interaction. *Procedia Computer Science, 3*: 1361–1367.

Dorneich, M. C., B. Passinger, C. Hamblin, C. Keinrath, J. Vasek, S. D. Whitlow, et al. (2011). The crew workload manager: An open-loop adaptive system design for next generation flight decks. *Proceedings of the Human Factors and Ergonomics Society Annual Meeting, 55*(1): 16–20.

Edwards, T. 2013. *Human Performance in Air Traffic Control*. Unpublished doctoral dissertation, University of Nottingham, Nottingham, United Kingdom.

Elkin-Frankston, S., B. K. Bracken, S. Irvin, and M. Jenkins. 2017. Are Behavioral Measures Useful for Detecting Cognitive Workload During Human-Computer Interaction? In *Advances in Intelligent Systems and Computing*, Volume 494, eds. T. Ahram, and W. Karwowski, 127–137. Cham: Springer.

Endsley, M. R. 1999. *Situation awareness and human error: Designing to support human performance. Proceedings of the High Consequence Systems Surety Conference.* Albuquerque, NM: Sandia National Laboratory.

Farmer, E., and A. Brownson. 2003. *Review of Workload Measurement, Analysis and Interpretation Methods*. Report prepared for EUROCONTROL INTEGRA programme (CARE-Integra-TRS-130-02-WP2). Retrieved August 16, 2020 from: http://citeseerx. ist.psu.edu/viewdoc/download?doi=10.1.1.121.3382&rep=rep1&type=pdf.

Faulstich, M. E., D. A. Williamson, S. J. McKenzie, E. G. Duchmann, K. M. Hutchinson, K., and D. C. Blouin. 1986. Temporal stability of psychophysiological responding: A comparative analysis of mental and physical stressors. *International Journal of Neuroscience, 30*(1–2): 65–72.

Feigh, K. M., M. C. Dorneich, and C. C. Hayes. 2012. Toward a Characterization of Adaptive Systems: A Framework for Researchers and System Designers. *Human Factors, 54*(6): 1008–1024.

Frank, G. R. 2007. Monitoring seated postural responses to assess cognitive state. Unpublished master's thesis, University of Pittsburgh, PA.

Fuchs, S., K. S. Hale, K. M. Stanney, C. Berka, D. Levendowski, and J. Juhnke. 2006. Physiological Sensors Cannot Effectively Drive System Mitigation Alone. In *Foundations of Augmented Cognition (2nd Ed.)*, eds. D. D. Schmorrow, K. M. Stanney, and L. M. Reeves, 193–200. Arlington, VA: Strategic Analysis, Inc.

Fuchs, S., K. S. Hale, K. M. Stanney, J. Juhnke, and D. D. Schmorrow. 2007. Enhancing Mitigation in Augmented Cognition. *Journal of Cognitive Engineering & Decision Making, 1*(3): 309–326.

Fuchs S., S. Hochgeschurz, A. Schmitz-Hübsch, and L. Thiele. 2020. Adapting Interaction to Address Critical User States of High Workload and Incorrect Attentional Focus – An Evaluation of Five Adaptation Strategies. In: *Augmented Cognition. Human Cognition and Behavior. HCII 2020. LNAI 12197*, eds. D. D. Schmorrow, and C. Fidopiastis, 335–352. Cham: Springer.

Fuchs, S., A. Schmitz-Hübsch, S. Hochgeschurz, L. Thiele, J. Schwarz, et al. 2019. Anwendungsorientierte Realisierung Adaptiver Mensch-Maschine-Interaktion für sicherheitskritische Systeme (ARAMIS). Final report for grant no. E/E4BX/HA031/ CF215. Wachtberg, Germany: Fraunhofer FKIE.

Fuchs, S., and J. Schwarz. 2017. Towards a dynamic selection and configuration of adaptation strategies in Augmented Cognition. In *Augmented Cognition 2017, Part II, LNAI 10285*, eds. D. D. Schmorrow, and C. M. Fidopiastis, 101–115. Cham: Springer.

Gagnon, J. F., S. Tremblay, D. Lafond, M. Rivest, and F. Couderc. 2014. *Sensor-Hub: A Real-Time Data Integration and Processing Nexus for Adaptive C2 Systems. Proceedings of the 6th IARIA Adaptive Conference*, 63–67. Venice, Italy.

Gateau, T., G. Durantin, F. Lancelot, S. Scannella, and F. Dehais. 2015. Real-Time State Estimation in a Flight Simulator Using fNIRS. *PLOS ONE* 10, Nr. 3: e0121279. doi:10.1371/journal.pone.0121279.

Ghergulescu, I., and C. H. Muntean. 2010. MoGAME: Motivation based Game Level Adaptation Mechanism. *Proceedings of the 10th Annual Irish Learning Technology Association Conference EdTech 2010*. Athlone, Ireland.

Ghergulescu, I., and C. H. Muntean. 2011. *Learner motivation assessment with <e-adventure> game platform. Proceedings of AACE E-LEARN-World Conference on E-Learning in Corporate, Government, Healthcare & Higher Education*, 1212–1221. Honolulu, Hawaii.

Goldberg, B. S., R. A. Sottilare, K. W. Brawner, and H. K. Holden. 2011. Predicting learner engagement during well-defined and ill-defined computer-based intercultural interactions. In: *International Conference on Affective Computing and Intelligent Interaction Part I, LNCS, vol. 6974*, eds. S. D'Mello, A. Graesser, B. Schuller, and J.-C. Martin, 538–547. Berlin, Heidelberg: Springer.

Grandt, M. 2004. *Erfassung und Bewertung der mentalen Beanspruchung mittels psychophysiologischer Messverfahren*. FKIE-Report Nr. 88. Wachtberg: Forschungsinstitut für Kommunikation, Informationsverarbeitung und Ergonomie.

Haarmann, A., W. Boucsein, and F. Schaefer. 2009. Combining electrodermal responses and cardiovascular measures for probing adaptive automation during simulated flight. *Applied Ergonomics*, 40(6): 1026–1040.

Hadley, G. A., L. J. Prinzel, F. G. Freeman, and P. J. Mikulka. 1999. Behavioral, subjective and psychophysiological correlates of various schedules of short-cycle automation. In *Automation Technology and Human Performance*, eds. M. W. Scerbo, and M. Mouloua, 139–143. Mahwah, NJ: Lawrence Erlbaum Assoc., Inc.

Hancock, P. A., and M. H. Chignell. 1987. Adaptive Control in Human-Machine Systems. In *Human Factors Psychology*, ed. P. A. Hancock, 305–345. Amsterdam: North-Holland.

Hancock, P. A., and S. F. Scallen. 1998. Allocating functions in human-machine systems. In *Viewing Psychology as a Whole: The Integrative Science of William N. Dember*, eds. R. R. Hoffman, M. F. Warm, and J. S. Sherrick. 509–539. Washington, DC: American Psychological Association.

Hancock, P. A., and W. B. Verwey. 1997. Fatigue, workload and adaptive driver systems. *Accident Analysis and Prevention*, 29(4): 495–506.

Hargutt, V. 2003. *Das Lidschlagverhalten als Indikator für Aufmerksamkeits- und Müdigkeitsprozesse bei Arbeitshandlungen*. Düsseldorf: VDI Verlag.

Hart, S. G., and L. E. Staveland. 1988. Development of a multi-dimensional workload rating scale: Results of empirical and theoretical research. In *Human mental workload*, eds. P. A. Hancock, and N. Meshkati, 139–183. Amsterdam, The Netherlands: Elsevier.

Hershkovitz, A., and R. Nachmias. 2011. Online persistence in higher education web-supported courses. *The Internet and Higher Education*, 14(2): 98–106.

Hincks, S. W., D. Afergan, and R. J. K. Jacob. 2016. Using fNIRS for Real-Time Cognitive Workload Assessment. In *Foundations of Augmented Cognition: Neuroergonomics and Operational Neuroscience*. AC 2016. Lecture Notes in Computer Science, vol. 9743, eds. D. Schmorrow, and C. Fidopiastis, 198–208. Cham: Springer.

Hirshfield, L., K. Chauncey, R. Gulotta et al. 2009. Combining electroencephalograph and functional near infrared spectroscopy to explore users' mental workload. In *Foundations of Augmented Cognition. Neuroergonomics and Operational Neuroscience, LNCS, vol. 5638*, eds. D. D. Schmorrow, I. V. Estabrooke, and M. Grootjen, 239–247. Berlin, Heidelberg: Springer.

Hockey, G. R. J. 2003. Operator functional state as a framework for the assessment of performance degradation. In *Operator Functional State*, eds. G. R. J. Hockey, A. W. K. Gaillard, and O. Burov, 8–23. Amsterdam: IOS Press.

Hoddes, E., V. Zarcone, H. Smythe, R. Phillips, and W. C. Dement. 1973. Quantification of sleepiness: a new approach. *Psychophysiology*,*10*(4): 431–436.

Hogervorst, M. A., A. M. Brouwer, and J. B. van Erp. 2014. Combining and comparing EEG, peripheral physiology and eye-related measures for the assessment of mental workload. *Frontiers in Neuroscience*, *8*(322). doi: 10.3389/fnins.2014.00322.

Hudlicka, E., and D. McNeese. 2002. Assessment of user affective and belief states for interface adaptation: Application to an Air Force pilot task. *User Modeling and User-Adapted Interaction*, *12*(1): 1–47.

Hursh, S. R., T. J. Balkin, J. C. Miller, and D. R. Eddy. 2004. The fatigue avoidance scheduling tool: Modeling to minimize the effects of fatigue on cognitive performance. *SAE Transactions*, *113*(1): 111–119.

Hurwitz, J. B., and D. J. Wheatley. 2002. Using driver performance measures to estimate workload. In *Proceedings of the Human Factors and Ergonomics Society Annual Meeting*, *46*(22): 1804–1808.

Inagaki, T. 2003. Adaptive Automation: Sharing and Trading of Control, In *Handbook of Cognitive Task Design*, ed. E. Hollnagel, 147–169. Mahwah, NJ: Lawrence Erlbaum Associates Publishers.

Ji, Q., Z. Zhu, and P. Lan. 2004. Real-time nonintrusive monitoring and prediction of driver fatigue, *Proceedings of IEEE Transactions on Vehicular Technology*, *53*(4): 1052–1068.

Kaber, D. B., L. J. Prinzel, M. C. Wright, and M. P. Clamann. 2002. *Workload-Matched Adaptive Automation Support of Air Traffic Controller Information Processing Stages*. Technical Paper NASA/TP-2002-211932. Hampton, Virginia: National Aeronautics and Space Administration, Langley Research Center.

Kaber, D. B., and M. C. Wright. 2003. Adaptive automation of stages of information processing and the relation to operator functional states. In *Operator Functional States*, vol 355, ed. G. R. J. Hockey, 204–223. IOS Press: NATO Science Series Sub Series I Life and Behavioral Science.

Kecklund, G., T. Åkerstedt, D. Sandberg, et al. 2007. *DROWSI - State of the art review of driver sleepiness*. IVSS project report.

Kramer, A. F. 1991. Physiological metrics of mental workload: A review of recent progress. In *Multiple Task Performance*, ed. D. Damos, 279–328. London: Taylor and Francis.

Lal, S. K., and A. Craig. 2001. A critical review of the psychophysiology of driver fatigue. *Biological psychology*, *55*(*3*): 173–194.

Lee, C. M., and S. S. Narayanan. 2005. Towards detecting emotions in spoken dialogs. *IEEE Transactions on Speech & Audio Processing*, *13*(2): 293–303.

Lin, C. T., L. W. Ko, I. F. Chung et al. 2006. Adaptive EEG-based alertness estimation system by using ICA-based fuzzy neural networks, *IEEE Transactions on Circuits and Systems I: Regular Papers*, *53*(11): 2469–2476.

Mackworth, N. H. 1948. The breakdown of vigilance during prolonged visual search. *Quarterly Journal of Experimental Psychology*, *1(1)*: 6–21.

Mahlke, S., and M. Minge. 2006. *Emotions and EMG measures of facial muscles in interactive contexts*, *Proceedings of CHI 2006*, Montreal, Canada.

Marshall, S. 2000. Method and Apparatus for Eye Tracking and Monitoring Pupil Dilation to Evaluate Cognitive Activity, U.S. Patent 6,090,051. Washington, DC: U.S. Patent and Trademark Office.

Marshall, S. 2002. *The index of cognitive activity: Measuring cognitive workload. Proceedings of the 2002 IEEE 7th Conference on Human Factors and Power Plants, 2002*, 7.5–7.9. New York: IEEE.

Mathan, S., D. Erdogmus, Y. Huang et al. 2008. *Rapid image analysis using neural signals. Proceedings of the twenty-sixth annual CHI conference extended abstracts on Human factors in computing systems - CHI '08*, 3309–3314. New York: ACM Press.

Mauss, I. B., and Robinson, M. D. (2009). Measures of emotion: A review. *Cognition and emotion*, *23*(2), 209–237.

May, J. F., and C. L. Baldwin. 2009. Driver fatigue: The importance of identifying causal factors of fatigue when considering detection and countermeasure technologies. *Transportation Research Part F: Traffic Psychology and Behaviour*, *12*(3): 218–224.

Meghanathan, R. N., C. van Leeuwen, and A. R. Nikolaev. 2014. Fixation duration surpasses pupil size as a measure of memory load in free viewing. *Frontiers in Human Neuroscience*, *8*, 1063. http://doi.org/10.3389/fnhum.2014.01063.

Miller, S. 2001. *Workload Measures. National Advanced Driving Simulator*. Oakland, IA: The University of Iowa.

Mulder, L. J. M. 1992. Measurement and analysis methods of heart rate and respiration for use in applied environments. *Biological Psychology*, *34*(2–3): 205–236.

Mulder, L. J. M., A. Kruizinga, A. Stuiver, I. Vernema, and P. Hoogeboom. 2004. Monitoring cardiovascular state changes in a simulated ambulance dispatch task for use in adaptive automation. In *Human Factors in Design*, eds. D. de Waard, K.A. Brookhuis, and C.M. Weikert, 161–175. Maastricht: Shaker Publishing.

Mulder, B., D. de Waard, P. Hoogeboom, L. Quispel, and A. Stuiver. 2008. Using Physiological Measures For Task Adaptation: Towards a Companion. In *Probing Experience: From Assessment of User Emotions and Behaviour to Development of Products*, eds. J. H. D. M. Westerink, M. Ouwerkerk, T. J. M. Overbeek, W. F. Pasveer, and B. de Ruyter, 221–234. Dordrecht, The Netherlands: Springer.

Nguyen, T., S. Ahn, H. Jang, S. C. Jun, and J. G. Kim. 2017. Utilization of a combined EEG/NIRS system to predict driver drowsiness. *Scientific Reports*, *7*, 43933. http://doi.org/10.1038/srep43933.

Noronha, H., R. Sol, and A. Vourvopoulos. 2013. Comparing the Levels of Frustration between an Eye-Tracker and a Mouse: A Pilot Study. In *Human Factors in Computing and Informatics*, 107–121. Berlin, Heidelberg: Springer.

Nwe, T. L., F. S. Wei, and L. C. De Silva. 2001. *Speech based emotion classification*. In *TENCON 2001. Proceedings of IEEE Region 10 International Conference on Electrical and Electronic Technology*, vol. 1, 297–301. New York: IEEE.

O'Donnell, R. D., and F. T. Eggemeier. 1986. Workload assessment methodology. In *Handbook of perception and human performance. Cognitive processes and performance*, eds. K. R. Boff, L. Kaufman, and J. P. Thomas, 1–49. New York: John Wiley & Sons.

Oken, B. S., M. C. Salinsky, and S. M. Elsas. 2006. Vigilance, alertness, or sustained attention: physiological basis and measurement. *Clinical Neurophysiology*, *117*(9): 1885–1901.

Parasuraman, R. 2003. Neuroergonomics: research and practice. *Theoretical Issues in Ergonomic Science*, *4*(1–2): 5–20.

Parasuraman, R., T. Bahri, J. E. Deaton, J. G. Morrison, and M. Barnes. 1992. *Theory and design of adaptive automation in aviation systems*. Progress Report NAWCADWAR-92033-60 under Contract No. N62269-90-0022-5931. Warminster, PA.

Parasuraman, R., M. Mouloua, and R. Molloy. 1996. Effects of adaptive task allocation on monitoring of automated systems. *Human Factors*, *38*(4): 665–679.

Parasuraman, R., T. B. Sheridan, and C. D. Wickens. 2000. A model of types and levels of human interaction with automation. *IEEE Transactions on Systems, Man, and Cybernetics – Part A: Systems and Humans*, *30*(3): 286–297.

Pope, A. T., E. H. Bogart, and D. S. Bartolome. 1995. Biocybernetic system evaluates indices of operator engagement in automated task. *Biological Psychology*, *40*(1–2): 187–195.

Prinzel III, L. J., A. T. Pope, F. G. Freeman, M. W. Scerbo, P. J. Mikulka, and L. J. Prinzel. 2001. *Empirical Analysis of EEG and ERPs for Psychophysiological Adaptive Task Allocation. NASA/TM-2001-211016*. Hampton, VA: National Aeronautics and Space Administration.

Roscoe, A. H. 1993. Heart rate as a psychophysiological measure for inflight workload assessment. *Ergonomics*, *36*(9): 1055–1062.

Rouse, W. B. 1988. Adaptive aiding for human/computer control. *Human Factors*, 30(4): 431–443.

Scallen, S. F., P. A. Hancock, and J. A. Duley. 1995. Pilot performance and preference for short cycles of automation in adaptive function allocation. *Applied Ergonomics*, *26*(6): 397–403.

Scerbo, M. W. 1996. Theoretical perspectives on adaptive automation. In, *Automation and Human Performance: Theory and Applications*, eds. R. Parasuraman, and M. Mouloua, 37–63. Hillsdale, NJ: Lawrence Erlbaum Associates, Inc.

Scerbo, M. W. 2001. Stress, workload and boredom in vigilance: a problem and an answer. In *Stress, Workload & Fatigue*, eds. P. Hancock, and P. Desmond, 267–278. Mahwah, NJ: Erlbaum.

Schleicher, R., N. Galley, S. Briest, and L. Galley. 2008. Blinks and saccades as indicators of fatigue in sleepiness warnings: Looking tired? *Ergonomics*, *51*(7): 982–1010.

Schwarz, J. 2019. Multifaktorielle Echtzeitdiagnose des Nutzerzustands in adaptiver Mensch-Maschine-Interaktion. PhD-Thesis Technische Universität Dortmund. doi: 10.17877/DE290R-20269

Schwarz, J., and S. Fuchs. 2016. Test-retest stability of eeg and eye tracking metrics as indicators of variations in user state—An analysis at a group and an individual level. In *Advances in Neuroergonomics and Cognitive Engineering*, eds. K. S. Hale, and K. M. Stanney, 145–156. Cham: Springer International Publishing.

Schwarz, J., and S. Fuchs. 2017. Multidimensional real-time assessment of user state and performance to trigger dynamic system adaptation. In *Augmented Cognition 2017. LNCS (LNAI)*, vol. 10285, eds. D. D. Schmorrow, and C. M. Fidopiastis, 383–398. Cham: Springer.

Schwarz, J., and S. Fuchs. 2018. *Validating a "Real-Time Assessment of Multidimensional User State" (RASMUS) for Adaptive Human-Computer Interaction*. In *Proceedings of the IEEE International Conference on Systems, Man and Cybernetics (SMC)*, Oct, 07–10, Miyazaki, Japan, 704–709.

Schwarz, J., S. Fuchs, and F. Flemisch. 2014. *Towards a more holistic view on user state assessment in adaptive human-computer interaction*. In *Proceedings of the IEEE International Conference on Systems, Man, and Cybernetics*, Oct 5–8, San Diego, CA, USA, 1247–1253.

Sciarini, L. W., and D. Nicholson. 2009. Assessing cognitive state with multiple physiological measures: A modular approach. In *Augmented Cognition, HCII 2009*, LNAI 5638, eds. D. D. Schmorrow, I. V. Estabrooke, and M. Grootjen, 533–542. Berlin: Springer.

Silvagni, S., L. Napoletano, I. Graziani, P. LeBlaye, and L. Rognin. 2015. Concept for Human Performance Envelope. Future Sky Safety P6 Human Performance Envelope Report D6.1. Retrieved August 16, 2020 from: https://www.futuresky-safety.eu/wp-content/uploads/2015/12/FSS_P6_DBL_D6.1-Concept-for-Human-Performance-Envelope_v2.0.pdf.

Son, J., and S. Park. 2011. *Cognitive workload estimation through lateral driving performance*. In *Proceedings of the 16th Asia Pacific Automotive Engineering Conference, 2011*. Warrendale, PA: SAE Technical Paper.

Staal, M. A. 2004. *Stress, Cognition, and Human Performance: A Literature Review and Conceptual Framework*. Hanover, MD: National Aeronautics & Space Administration.

Stanney, K. M., D. D. Schmorrow, M. Johnston et al. 2009. Augmented Cognition: An Overview. In *Reviews of Human Factors and Ergonomics*, ed. F.T. Durso, vol. 5, 195–224. Santa Monica, CA: HFES.

Steinhauser, N. B., D. Pavlas, and P. A. Hancock. 2009. Design principles for adaptive automation and aiding. *Ergonomics in Design*, *17*(2): 6–10.

Strait, M., and M. Scheutz. 2014. What we can and cannot (yet) do with functional near infrared spectroscopy. *Frontiers in Neuroscience, 8*, 117. http://doi.org/10.3389/fnins.2014.00117

Stuiber, G. 2006. Studie zur Untersuchung der Auswirkungen von Koffein auf den Pupillographischen Schläfrigkeitstest bei gesunden Probanden. Unpublished doctoral dissertation, Universität Tübingen, Tübingen, Germany. Retrieved 16 August 2020 from: http://nbn-resolving.de/urn:nbn:de:bsz:21-opus-25246

Ting, C. H., M. Mahfouf, D. A. Linkens et al. 2008. Real-time adaptive automation for performance enhancement of operators in a human-machine system. In *16th Mediterranean Conference on Control and Automation*, 552–557. New York: IEEE.

Tomarken, A. J. 1995. A psychometric perspective on psychophysiological measures. *Psychological Assessment, 7*(3): 387–395.

Touré-Tillery, M., and A. Fishbach. 2014. How to measure motivation: A guide for the experimental social psychologist. *Social and Personality Psychology Compass, 8*(7): 328–341.

Tremoulet, P., J. Barton, and P. Craven. 2005. *DARPA Improving Warfighter Information Intake Under Stress - Augmented Cognition Phase 3*. Concept Validation Experiment (CVE) Analysis Report for the Lockheed Martin ATL Team prepared under contract NBCHC030032. Arlington, VA: DARPA/IPTO.

Uhlig, S. 2018. Heart Rate Variability: What remains at the end of the day? Doctoral dissertation, TU Chemnitz. Retrieved August 16 2020 from: http://nbn-resolving.de/urn:nbn:de:bsz:ch1-qucosa-233101.

Van Orden, K. F., W. Limbert, S. Makeig, and T. Jung. 2001. Eye activity correlates of workload during a visuospatial memory task. *Human Factors, 43*(1): 111–121.

Veltman, J. A., and A. W. K. Gaillard. 1998. Physiological workload reactions to increasing levels of task difficulty. *Ergonomics, 41*(5): 656–669.

Veltman, J. A., and C. Jansen. 2003. Differentiation of Mental Effort measures: Consequences for Adaptive Automation. In *Operator Functional State*, eds. G. R. J. Hockey, A. W. K. Gaillard, and O. Burov, 249–259. Amsterdam: IOS Press.

Veltman, J. A., and C. Jansen. 2004. The adaptive operator. In *Human Performance, Situation Awareness and Automation Technology*, eds. D.A. Vincenzi, M. Mouloua, and P.A. Hancock, Vol. 2, 7–10. Mahwah, NJ: Lawrence Erlbaum Associates.

Veltman, J. A., and C. Jansen. 2006. *The role of operator state assessment in adaptive automation*. TNO Report TNO-DV3 2005 A245. Soesterberg, Netherlands: TNO Report.

Wang, L., V. G. Duffy, and Y. Du. 2007. A composite measure for the evaluation of mental workload. In *Digital Human Modeling, HCII 2007, LNCS 4561*, eds. V.G. Duffy, 460–466.

Whitlow, S., and Hayes, C. C. (2012). Considering Etiquette in the Design of an Adaptive System. *Journal of Cognitive Engineering and Decision Making, 6*(2), 243–265.

Wickens, C. D. 2005. *Attentional tunneling and task management*. In *Proceedings of the 13th International Symposium on Aviation Psychology*, 620–625.

Wilson, G. F., W. Fraser, M. Beaumont, et al. 2004. *Operator functional state assessment*. (NATO RTO Publication RTO-TR-HFM-104). Neuilly sur Seine: NATO Research and Technology Organization.

Wilson, G. F., and C. A. Russell. 2003. Operator functional state classification using multiple psychophysiological features in an air traffic control task. *Human Factors, 45*(3): 381–389.

Wilson, G. F., and C. A. Russell. 2006. Psychophysiologically Versus Task Determined Adaptive Aiding Accomplishment. In *Foundations of Augmented Cognition*, eds. D. D. Schmorrow, K. M. Stanney, and L. M. Reeves, 201–207. Arlington, VA: Strategic Analysis, Inc.

Woolf, B., W. Burleson, I. Arroyo, T. Dragon, D. Cooper, and R. Picard. 2009. Affect-aware tutors: Recognising and responding to student affect. *International Journal of Learning Technology, 4*(3–4):129–164.

Wright, N. A., B. M. Stone, T. J. Horberry, and N. Reed. 2007. *A Review of In-vehicle Sleepiness Detection Devices*. Published project report 157. Wokingham, UK: TRL Limited.

Zijlstra, F. R. H. 1993. *Efficiency in Work Behaviour: A Design Approach for Modern Tools*. Delft: Delft University Press.

4 Agent Transparency

Jessie Y.C. Chen

CONTENTS

4.1 INTRODUCTION

Humans are increasingly working with intelligent agents—from robots to complex networked systems—as collaborators rather than tool users, due to the tremendous advancements in artificial intelligence (AI) in recent years. In order for human–agent teaming and communications to work effectively, agent transparency is a critical component that needs to be implemented in the joint human–agent systems. Indeed, the European Union recently announced the top three requirements for "high-risk" AI-based systems (e.g. health, policing, transport): transparency, traceability, and human control (European Commission 2020). One of the examples of such AI application is using machine learning (ML) to classify diseases such as macular degeneration (Lu 2020). These applications have shown significant promise and, in cases, have shown higher accuracy rates than humans'; attention is increasingly focused on making the output transparent to medical personnel to maximize the human–machine team performance (Lu 2020).

A recent special issue on Agent and System Transparency published in *IEEE Transactions on Human–Machine Systems* represents a major milestone for transparency research (Chen et al. 2020). High-profile research programs such as DARPA's eXplainable AI (XAI) also show Government agencies' investment in transparency-related research (Gunning and Aha 2019; Miller 2019). Tech industry, where AI is a critical capability, is keenly aware of the importance of system transparency to the overall user–system interaction effectiveness. For example, according to Xavier Amatriain, who developed the recommender system for Netflix, system explanations to the users (about the reasons of recommendations) were often more effective than tweaking the algorithms (Amatriain 2016).

Agent transparency has significant implications for a wide variety of human–agent interaction environments where joint decision-making is required. Indeed, as systems become more intelligent and are capable of complex decision-making, it becomes increasingly important for humans to understand the reasoning process behind the agent's output in order to provide input when necessary. For example, the agent may indicate that the current plan is based on considering constraints x, y, and z. Human operators, based on their understanding of the mission requirements and information not necessarily accessible to the agent (e.g. intelligence reports), may instruct the agent to ignore z and focus only on the other two constraints. Without the agent's explanation of its reasoning process, the human operators may not realize that the agent's planning is suboptimal due to its consideration of an unimportant constraint. However, transparency for an AI-based system poses a challenge, as explanations may not be easily generated; yet, without proper transparency—at least high-level explanations—it may be difficult for the human to provide input (based on information that only the human has) to the system. Furthermore, predictability may be an issue for AI systems as they continue to evolve. Without proper transparency, the human may find it difficult to calibrate his/her trust in the system.

The Section 4.2 discusses the impact of agent transparency on human operator performance, situation awareness, trust calibration, and workload. Empirical findings on individual and cultural differences in human interaction with transparent systems will also be briefly summarized. The subsequent sections review a transparency framework called Situation awareness-based Agent Transparency (SAT) and applications of this framework to human–machine interface designs in a variety of human–autonomy interaction environments (i.e. small robot, multiagent systems, and robotic swarms). The final section of this chapter will discuss current challenges and suggest future research directions.

4.2 OPERATOR PERFORMANCE ISSUES

4.2.1 Operator Performance and Workload

The effects of agent transparency on operator performance have been investigated in multiple studies (see Bhaskara et al. 2020 for a review). In general, the findings suggest that agent transparency benefits operator performance (e.g. correct usage of agent recommendations and greater operator situation awareness) in a diverse human–agent interaction context. For example, Mercado et al. (2016) and Stowers et al. (2020) found that when interacting with an intelligent planning agent in a multirobot management tasking environment, human operators performed better (i.e. deciding whether to accept the agent's recommendations) with a more transparent agent. In other words, human operators' calibration of their trust was better when dealing with a more transparent agent and knowing when its input should be rejected. Similarly, Helldin et al. (2014) found that transparency of an automated target classifier (in terms of sensor accuracy and uncertainty information) benefitted fighter pilots' target classification performance. In another study, Zhou et al. (2016) found that in the context of a human collaborating with a ML-based agent to predict system failures based on historical data, agent transparency (visualizing its internal algorithmic processes) had a positive impact on human–agent interaction—the operators understood the agent's analysis process better and found the agent's recommendations more convincing. Multiple studies have demonstrated the benefits of presenting machine agents' uncertainties on human–agent team performance (Bass, Baumgart and Shepley 2013; Beller, Heesen and Vollrath 2013; Mercado et al. 2016). However, these benefits might be mitigated somewhat by the information processing requirement associated with uncertainty information potentially slowing down operators' decision-making process by a few seconds (Stowers et al. 2020).

While the benefits of agent transparency on operator performance have been well documented, there have been inconsistent findings regarding the effects of transparency on operator workload. Helldin et al. (2014) found that there were costs associated with greater system transparency in terms of increased operator workload and, additionally, decision response times could be impacted due to the large amount of information to be processed associated with transparent interfaces. It is also worth noting that information about uncertainty could increase the operator's workload. Kunze et al. (2019) found that during human interaction with automated driving, transparency information conveying uncertainties benefited the operator's overall task performance (including trust calibration and situation awareness) but significantly increased operator workload. Conversely, however, the same result of increased operator workload was not observed in other transparency studies (Mercado et al. 2016; Selkowitz et al. 2017; Stowers et al. 2020) except the minor increase of response times of a few seconds reported in Stowers et al.

4.2.2 OPERATOR TRUST AND PERCEPTIONS OF TRANSPARENT AGENTS

Agent transparency also impacts humans' perceptions of the agent. Several empirical studies have investigated the effects of agent transparency on operator (subjective) trust, and the findings have been largely consistent—higher levels of operator trust in the intelligent agent as it became more transparent (Helldin et al. 2014; Mercado et al. 2016; Vered et al. 2020; Selkowitz et al. 2017; Stowers et al. 2020; Zhou and Chen 2015). However, the elevated trust associated with increased transparency could be degraded when the agent conveyed uncertainty, particularly in the context of recommending courses of action (Stowers et al. 2020). In multiple studies, research participants reported higher levels of perceived intelligence and human-likeness when the agent was more transparent (Mercado et al. 2016; Roth et al. 2020; Wohleber et al. 2017). However, the ways the transparency information is conveyed could affect the perceptions. For example, Wohleber et al. manipulated the framing of the agent recommendations (as compliments vs. critiques) and found that the critical agent was perceived to be more capable in its tasks compared with a more complimentary agent.

4.2.3 INDIVIDUAL AND CULTURAL DIFFERENCES

The effects of agent transparency have been examined in the contexts of individual and cultural differences. With regard to the individual aspects, Matthews et al. (2020) investigate the relationships between humans' mental models of robots' analytic performance (whether the robot's analysis is based on physics- or psychology-related info in the experimental threat-based scenarios) and individual differences in attitudes toward robots in general (measured by the perfect automation schema and the negative attitudes toward robots scales). The authors conclude that, given human biases (e.g. unreasonable expectations of robot capability or negative attitudes toward humanlike robots) and their potential impact on operator trust calibration and SA, transparency information should be contextual and personalized. More specifically, transparency content should be compatible with the operator's mental model (e.g. the robot as a tool or a teammate) in order to avoid over- or under-trust. Furthermore, when possible, user interfaces should be personalized to highlight the robot's data-analytic capabilities or its social (e.g. humanlike) aspects. Based on the results, the authors provide transparent interface design suggestions based on the SAT framework, which will be discussed in Section 4.3.

At the cultural level, Chien et al. (2020) examine human–autonomy interaction with different degrees of automation and transparency in three distinct cultural backgrounds (based on the Cultural Syndromes Theory): United States (Dignity), Taiwan (Face), and Turkey (Honor). The research participants' task was to work with an intelligent planning agent with different degrees of automation to manage a team of five unmanned aerial vehicles (UAVs) and, simultaneously, identify/attack hostile targets and reroute the UAVs when necessary. The simulation-based experimental results show that agent transparency had an impact on operator's interaction with the planning agent (i.e. compliance with agent's recommendations), but the effects of agent transparency were significantly influenced by participants' culture. For example, Face culture participants had a higher tendency to accept recommendations from an opaque agent than did participants from the other two cultural backgrounds. These results suggest that when transitioning autonomy technologies from one culture to another, user interface modifications and training interventions might be required due to the effects of cultural differences on system reliance related to agent transparency.

4.3   SITUATION AWARENESS-BASED AGENT TRANSPARENCY

Chen and her colleagues define agent transparency as "descriptive quality of an interface pertaining to its abilities to afford an operator's comprehension about an intelligent agent's intent, performance, future plans, and reasoning process" (p. 2, Chen et al. 2014). One of the transparency frameworks that has been developed and investigated in a variety of simulation environments is SAT (Bhaskara et al. 2020; Chen et al. 2018; Roth et al. 2020; Stowers et al. 2020; Wright et al. 2020), which is based on Endsley's (1995) model of situation awareness (three levels of SA—perception, comprehension, and projection), Lee's (2012) model incorporating the 3Ps (purpose, process, and performance) of the system (Lee and See 2004), and the BDI (Beliefs, Desires, Intentions) Agent Framework (Rao and Georgeff 1995) (Figure 4.1).

The SAT model is comprises three independent levels, each of which describes the information an agent needs to convey to the human in order to ensure transparency and to support shared situation understanding within the human–agent team

Situation awareness-based Agent Transparency

Level 1: Goals & Actions
Agent's current status/actions/plans
- Purpose: Desire (Goal selection)
- Process: Intentions (Planning/Execution); Progress
- Performance
- Perception (Environment/Teammates)

Level 2: Reasoning
Agent's reasoning process
- Reasoning process (Belief/Purpose)
- Motivations
  - Environmental and other constraints/affordances

Level 3: Projections
Agent's projections/predictions; uncertainty
- Projection of future outcomes
- Uncertainty and potential limitations; Likelihood of success/failure
- History of Performance

FIGURE 4.1   Situation awareness-based Agent Transparency (Chen et al. 2018).

(Chen et al. 2014; Endsley 1995). At the first SAT level (SAT 1), the agent provides the human with the basic information about the agent's current state and goals, intentions, and plans. This level of information assists the human's *perception* of the agent's current actions and plans. At the second level (SAT 2), the agent conveys its reasoning process as well as the constraints/affordances that the agent considers when planning its actions. In this way, SAT 2 supports the human's *comprehension* of the agent's current plans and actions. At the third SAT level (SAT 3), the agent provides the operator with information regarding its projection of future states, predicted outcomes of current actions and plans (e.g. likelihood of success/failure), and any uncertainty associated with the aforementioned projections. Thus, SAT 3 information assists with the human's *projection* of future outcomes.

The SAT framework could be used to not only identify the information requirements needed to facilitate effective human–agent joint decision-making and shared situation understanding but also to address operator trust and trust calibration. Particularly, SAT 3 encompasses transparency with regard to uncertainty, given that projection is predicated on many factors whose outcomes might not be precisely known (Bass, Baumgart, and Shepley 2013; Chen et al. 2014; Helldin 2014; Mercado et al. 2016; Meyer and Lee 2013). Previous research findings suggest that uncertainty information not only benefits human–agent team performance but also human's perceived trustworthiness of the agent (Beller, Heesen, and Vollrath 2013; McGuirl and Sarter 2006; Mercado et al. 2016). Several human–machine interfaces have been designed based on the SAT framework; the following section briefly summarizes these projects and their associated human factor studies to illustrate the effects of agent transparency on operator performance and trust calibration in human–agent teaming task environments.

4.4   IMPLEMENTATION OF TRANSPARENCY IN COMPLEX HUMAN–AGENT SYSTEMS

The SAT framework has been successfully applied to the design of the Transparency Module of the human–machine interface for several projects, with diverse contexts from human interaction with a small ground robot, to multiagent management systems, to human–swarm interaction. Two of these projects are part of the U.S. Department of Defense (DoD) Autonomous Research Pilot Initiative program: IMPACT (a planning agent in the context of multirobot management; Mercado et al. 2016; Stowers et al. 2020) and Autonomous Squad Member (ASM) (a small ground robot application; Selkowitz et al. 2017; Wright et al. 2020), both of which are discussed in detail in this section. SAT-based interface was also implemented for a workload-adaptive cognitive agent in a multirobot management context to support a military helicopter crew (Roth et al. 2020). Similar to findings Section 4.3, transparency significantly enhanced the human teammates' situation awareness and task performance. Additionally, the transparent agent was perceived as more humanlike than the opaque agent.

It is worth noting that the utility of SAT (or any transparency framework) depends on the effectiveness of how the information is presented and not just on what information is presented. Since transparency inherently involves *more* information to be included and conveyed to the user, it is particularly imperative that sound user interface design principles be followed so the user is not overwhelmed by the additional information. In high-tempo tasking environments where large amounts of information need to be delivered quickly, interface design strategies such as visualizations could be useful (see Cha et al. 2019 for a detailed review). Recent efforts have also started to examine ways to promote bidirectional/mutual transparency—not only of the robot (to the human) but also of the human (to the robot) (Chen et al. 2018; Lyons 2013; Wynne and Lyons 2018). For example, Lakhmani (2019) tried to tackle the bidirectional transparency issue in a simulated human–robot collaboration study, in which he manipulated both the robot transparency level (about itself only vs about both itself and the human teammate) and its communication pattern (conveying information to the human or "push" only vs both "push" and "pull," i.e. soliciting information from the human). Participants reported that they found robots that both pushed and pulled information and, compared with the less interactive robots, were perceived to be more animate, likeable, and intelligent, although interacting with those robots had a negative impact on the participants' concurrent task performance. This study illustrates the tremendous challenge of designing effective human–agent interfaces to promote bidirectional transparency in highly complex and dynamic tasking situations.

4.4.1   HUMAN INTERACTION WITH A SMALL ROBOT

The objective of the ASM project was to develop a small ground robot that could be embedded in a dismounted infantry squad (e.g. for load carrying purposes) and with capabilities to understand its environment and recognize anomalous situations and formulate new goals when necessary (Gillespie et al. 2015; Selkowitz et al. 2017;

Wright et al. 2020). One of the research aims of the ASM project is transparent human–robot interface in terms of supporting human teammates' awareness of the ASM's current actions, plans, perceptions of the environment/squad, its reasoning, and its projected outcomes (Selkowitz et al. 2017; Wright et al. 2020). The ASM interface not only conveys information about itself and its environment but also indicates its perceived actions of its human teammates to support the human's awareness. The ASM's user interface was designed based on the SAT framework and featured an "at a glance" transparency module (upper left corner of Figure 4.2), where iconographic representations of the agent's plans, motivator, and projected outcomes are used to promote transparent interaction with the agent (Selkowitz et al. 2017). The icons were integrated using Ecological Interface Design principles (Bennett and Flach, 2013) conveying the ASM's current actions and plans (SAT Level 1), motivators (SAT Level 2), projections of future outcomes, and uncertainties (SAT Level 3). A series of simulation-based studies on the ASM's transparent interface have been conducted to examine its effects on human teammates' SA, workload, and perceptions of the ASM such as trustworthiness, intelligence, and anthropomorphism (Selkowitz et al. 2017; Wright et al. 2020). The results of Selkowitz et al. showed that research participants had greater SA and trust in the ASM when it was more transparent. The additional information required for higher levels of transparency did increase participants' fixations on the transparency module but no differences in subjective workload were detected among different levels of transparency conditions. Wright et al. modified the

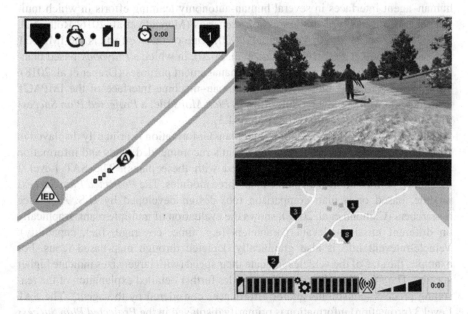

FIGURE 4.2 Autonomous Squad Member human–robot interface. "At-a-glance" transparency module based on the SAT framework is shown in the upper left corner. The transparency module (top row providing high-level information and the second row presenting further details) indicates the ASM's current status (left), top motivator (center), and predicted outcomes (right).

transparency "at-a-glance" module to provide two layers of SAT-based information, and additionally, the reliability of the ASM was manipulated. The results showed that participants' trust and perceptions of the robot were largely influenced by its reliability rather than transparency.

The SAT framework has also been implemented in a context where the human–robot communication is based on dialogue and natural language processing (NLP) to enhance the robot's shared mental model and Theory of Mind capabilities (Pynadath et al. 2018; Wang, et al. 2016; Wang et al. 2017). In a series of experiments, Wang and her colleagues examined the efficacy of robot transparency (based on the SAT framework) on human–robot teaming to conduct military reconnaissance tasks. In a typical experimental scenario, the robot would provide recommendations to the human participant about whether to don an item of protective gear based on the robot's sensor input (e.g. chem-bio, camera, or microphone). The level of transparency of the explanations was manipulated based on the SAT framework. The results show that robot transparency (about the reasoning process behind its recommendations) has a positive effect on the human–robot team performance especially when robots are not perfectly reliable, as it enables human teammates to identify robots' incorrect input or assumptions.

4.4.2 Multiagent Systems

The SAT framework (or its modified version) has been used to guide the designs of human–agent interfaces in several human-autonomy teaming efforts in which multiple agents are involved in the mission scenarios (Mercado et al. 2016; Roth et al. 2020; Stowers et al. 2020; Vered et al. 2020). One of these efforts is the IMPACT project (Mercado et al. 2016; Stowers et al. 2020), in which a *Playbook*-based planning agent was developed for multirobot management purposes (Draper et al. 2018). To convey SAT-based information, the human–machine interface of the IMPACT testbed contains three key modules/tiles: a *Plan Maps* tile, a *Projected Plan Success* tile, and a *Plan Specifications* tile (Figure 4.3).

The SAT Level 1 (current actions and plans) information is primarily displayed in the *Plan Maps* area (e.g. the planning agent's recommended plans and information about the autonomous vehicles associated with these plans). The SAT Level 2 (reasoning) information is available in all three modules. The *Projected Plan Success* module, based on a plan comparison tool design developed by U.S. Air Force researchers (Calhoun et al. 2018), shows the evaluation of multiple plans graphically on different mission-relevant parameters (e.g. time, coverage, fuel, capability). Vehicle-relevant info is also graphically depicted through map-based icons. For example, the size of the vehicles indicate their speed (with larger sizes indicate higher speeds). The *Plan Specifications* tile provides further detailed explanation of the reasoning process (constraints and/or affordances considered by the agent). The SAT Level 3 (projection) information is primarily displayed in the *Projected Plan Success* module, with the predicted outcomes of each plan graphically shown along the mission-relevant parameters and uncertainty indicated with hollow circles instead of filled ones. Further details are also available on the *Plan Specifications* tile (e.g. assumptions, uncertainties about certain aspects of the tasking environment).

FIGURE 4.3 IMPACT simulation experimental environment. Transparency designs based on SAT: "projected plan success" (Mission Metrics in lower left) (unfilled circles convey uncertainty), "plan specifications" text box (darker text conveys uncertainty associated with reasoning), and the map areas (vehicles/items with uncertain status/info are translucent).

In a series of simulation-based experiments using the IMPACT testbed, Mercado et al. (2016) and Stowers et al. (2020) investigated the effects of agent transparency (based on the SAT framework) on human-autonomy teaming in multiagent management mission contexts. In the experimental scenarios, participants' task was to decide whether to accept the planning agent's recommended plan (Plan A) or a backup plan (Plan B). Occasionally (3 out of 8 times), Plan B would be a better plan due to new information (e.g. newly developed environmental conditions, intelligence, or commander's instructions) to which the agent did not have access. Transparency of the agent's communication (based on SAT) was manipulated in order to test its effect on the operator's decision-making performance (i.e. selecting Plan A when appropriate and rejecting Plan A when Plan B was better). In Mercado et al. (2016), the results show that participants performed better with a more transparent agent without increased workload or response time. The participants calibrated their trust in the agent more appropriately (complying with the agent's recommendations when it was correct and rejecting the recommendations when it was incorrect) and rated the agent as more trustworthy when the agent was more transparent. In Stowers et al. (2020), the results show a similar pattern of benefits of transparency for supporting operator decision-making (i.e. proper trust calibration) and subjective trust in the agent without increasing workload. However, unlike the Mercado et al. study, participants' response time increased by a few seconds in the most transparent condition.

In another study, Roth et al. (2020) investigates the transparency issue in a military multiagent manned–unmanned teaming context, in which a helicopter crew collaborates with a cognitive agent to manage multiple UAVs. The cognitive agent is capable to assist the pilot adaptively based on his/her workload levels, and the agent's human–machine interface was designed based on the SAT framework to support the pilot's situation awareness and comprehension of the agent's interventions. A simulation-based human factors experiment showed that transparent interfaces enhanced participants' performance and situation awareness, although the workload results were less conclusive. Participants perceived the agent as more humanlike when it was more transparent, particularly when performing tasks such as mission management.

A framework similar to SAT (but focused only on the agent's reasoning process), the Endsley-based Transparency Model, was developed for a multirobot management system (Vered et al. 2020). In a simulation-based experiment with military surveillance scenarios, Vered et al. demonstrated that the way the transparency information was delivered to the operator impacted his/her performance and trust in the agent. In the experiment, when the information was introduced interactively (i.e. demand-driven), compared with when it was sequential, participants' performance (particularly speed of the decision-making process) was significantly better. The subjective trust in the agent was also significantly higher in the interactive group. Based on these results, the authors recommend interactive transparent interfaces as a promising design strategy for human-autonomy teaming in multirobot management environments.

4.4.3   Human–Swarm Interaction

Robotic swarms–typically large numbers (e.g. more than 50) of homogeneous miniature robotic agents–have been applied to a wide range of operational settings, such as search and rescue, military surveillance and reconnaissance, environmental monitoring and tracking, agriculture, and space exploration (Kolling et al. 2016). Robotic swarms, although each with only simple and limited capabilities, can exhibit emergent behaviors (e.g. flocking, rendezvous, and dispersion) by coordinating with one another via simple local communications and control laws (e.g. attraction, repulsion, and orientation) and can achieve substantial robustness and resiliency due to the responsibilities being distributed among the members of the swarm (Couzin et al. 2002; Roundtree et al. 2019; Nam et al. 2020). Given these characteristics, human–swarm interaction has to deal with challenges associated with managing large numbers of agents and uncertainties due to latency and asynchrony between human input and swarm responses, which affects effective assessments of the state and dynamics of the swarm and predictions of emergent behaviors based on human input.

Roundtree et al. (2019) identify some of the key challenges associated with applying transparency design principles based on non-swarm interfaces. Specifically, Roundtree et al. discuss in great detail how human–swarm interfaces could be designed to achieve the three levels of SAT. In a follow-on human–swarm interaction study, Roundtree et al. (2020) investigated the effects of two different visualization design methods (collective vs. individual-agent) in experimental scenarios where the participants managed a team of 200 agents. They found that the collective design achieved better transparency and supported human participants' SA better than the individual-agent design. Based on the experimental results, the authors put forward a list of design guidelines on transparent human–swarm interface visualizations. Particularly, the authors caution that operator individual differences (e.g. spatial abilities and cognitive capabilities) should be taken into consideration.

4.5 CHALLENGES AND FUTURE RESEARCH

This chapter reviews theories and techniques of agent transparency and their associated findings in human–agent interaction research. There remain some key challenges that will require further research and development to advance the capabilities of transparent interfaces. First of all, real-time and dynamic generation of transparency content requires interdisciplinary research (e.g. AI, robotics, human factors, linguistics) and remains a tremendous challenge. Research efforts in the United States such as the SUCCESS MURI program funded by the U.S. Navy (Han et al. 2019) represent a promising effort toward that goal. Information architectures of agent transparency should be developed to support real-time bidirectional transparency between human and machine agents. Ultimately, a well-designed transparency interface should not only convey the agent about itself and its understanding of the human's task and their shared tasks but also support effective "tweaking" by the human, who can thereby correct the agent's understanding or to provide instructions/information that the agent is not aware of. The SAT model and the bidirectional version of it (Chen et al. 2018) represent a useful framework that could potentially be further developed for various human–agent teaming contexts.

REFERENCES

Amatriain, X. 2016. *Past, present, and future of recommender systems: An industry perspective. Keynote speech at the International Conference on Intelligent User Interfaces*, Sonoma, CA, March 7, 2016.

Bass, E. J., Baumgart, L. A., and Shepley, K. K. 2013. The effect of information analysis automation display content on human judgment performance in noisy environments. *Journal of Cognitive Engineering and Decision Making* 7: 49–65.

Beller, J., Heesen, M., and Vollrath, M. 2013. Improving the driver-automation interaction: An approach using automation uncertainty. *Human Factors: The Journal of the Human Factors and Ergonomics Society* 55: 1130–1141.

Bennett, K. B., and Flach, J. M. 2013. Configural and pictorial displays. In *The Oxford Handbook of Cognitive Engineering*, eds. J. D. Lee & A. Kirlik, 517–533. New York: Oxford University Press.

Bhaskara, A., Skinner, M., and Loft, S. 2020. Agent transparency: A review of current theory and evidence. *IEEE Transactions on Human-Machine Systems* 50: 215–224.

Calhoun, G. L., Ruff, H., Behymer, K. J., et al. 2018. Human–autonomy teaming interface design considerations for multi-unmanned vehicle control. *Theoretical Issues in Ergonomics Science* 19: 321–352.

Cha, J., Barnes, M. J., and Chen, J. Y. C. 2019. *Visualization Techniques for Transparent Human Agent Interface Design* (Technical Report ARL-TR-8674). Aberdeen Proving Ground, MD: US Army Research Laboratory.

Chen, J. Y. C., Procci, K., Boyce, M., et al. 2014. *Situation Awareness-based Agent Transparency (ARL-TR-6905)*. Aberdeen Proving Ground, MD: U.S. Army Research Laboratory.

Chen, J. Y. C., Flemisch, F. O., Lyons J. B., et al. 2020. Guest editorial: Agent and system transparency. *IEEE Transactions on Human-Machine Systems* 50: 189–193.

Chen, J. Y. C., Lakhmani, S., Stowers, K., et al. 2018. Situation awareness-based agent transparency and human-autonomy teaming effectiveness. *Theoretical Issues in Ergonomics Science* 19: 259–282.

Chien, S., Lewis, M., Sycara, K., et al. 2020. Influence of culture, transparency, trust, and degree of automation on automation use. *IEEE Transactions on Human-Machine Systems* 50: 205–214.

Couzin, I. D., Krause, J., James, R., et al. 2002. Collective memory and spatial sorting in animal groups. *Journal of Theoretical Biology* 218: 1–11.

Draper, M., Rowe, A., Douglass, S., et al. 2018. *Realizing autonomy via intelligent adaptive hybrid control: Adaptable autonomy for achieving UxV RSTA team decision superiority (also known as Intelligent Multi-UxV Planner with Adaptive Collaborative/Control Technologies (IMPACT))* (Tech Rep. AFRL-RH-WP-TR-2018-0005). Wright-Patterson Air Force Base, OH: U.S. Air Force Research Laboratory.

Endsley, M. R. 1995. Toward a theory of situation awareness in dynamic systems. *Human Factors: The Journal of the Human Factors and Ergonomics Society* 37: 32–64.

European Commission. 2020. White Paper on Artificial Intelligence: A European approach to excellence and trust (Retrieved 10 June 2020, https://ec.europa.eu/info/files/white-paper-artificial-intelligence-european-approach-excellence-and-trust_en).

Gillespie, K., Molineaux, M., Floyd, M. W., et al. 2015. Goal reasoning for an autonomous squad member. *Proceedings of the Third Annual Conference on Advances in Cognitive Systems: Workshop on Goal Reasoning*. Atlanta, GA: Cognitive Systems Foundation.

Gunning, D., and Aha, D. 2019. DARPA's explainable artificial intelligence (XAI) program. *AI Magazine* 40: 44–58.

Han, Z., Allspaw, J., Norton, A., et al. 2019. *Towards a robot explanation system: A survey and our approach to state summarization, storage and querying, and human interface. Proceedings of the AI-HRI Symposium at the Association for the Advancement of Artificial Intelligent (AAAI) Fall Symposium Series (FSS) 2019.*

Helldin, T. 2014. Transparency for future semi-automated systems. PhD diss., Orebro University.

Helldin, T., Ohlander, U., Falkman, G., et al. 2014. Transparency of automated combat classification. In *Engineering Psychology and Cognitive Ergonomics. EPCE 2014. Lecture Notes in Computer Science*, vol. 8532, ed. D. Harris. Cham: Springer.

Kolling, A., Walker, P., Chakraborty, N., et al. 2016. Human interaction with robot swarms: A survey. *IEEE Transactions on Human-Machine Systems* 46: 9–26.

Kunze, A., Summerskill, S. J., Marshall, R., et al. 2019. Automation transparency: implications of uncertainty communication for human-automation interaction and interfaces. *Ergonomics* 62: 345–360.

Lakhmani, S. G. 2019. Transparency and communication patterns in human-robot teaming. PhD diss., University of Central Florida.

Lee, J. D. 2012. Trust, trustworthiness, and trustability. *Presented at the Workshop on Human Machine Trust for Robust Autonomous Systems*. Ocala, FL (2012, Jan-Feb).

Lee, J. D., and See, K. A. 2004. Trust in automation: Designing for appropriate reliance. *Human Factors: The Journal of the Human Factors and Ergonomics Society* 46: 50–80.

Lu, Z. 2020. *The role of AI in advancing health. Proceedings of the American Association for the Advancement of Science (AAAS) Annual Meeting – Symposium "The Human Impacts of AI."*

Lyons, J. B. 2013. *Being transparent about transparency: A model for human-robot interaction. Proceedings of the AI-HRI Symposium at the Association for the Advancement of Artificial Intelligent (AAAI) Spring Symposium Series* (pp. 48–53).

Matthews, G., Lin, J., Panganiban, A. R., et al. 2020. Individual differences in trust in autonomous robots: Implications for transparency. *IEEE Transactions on Human-Machine Systems* 50: 234–244.

McGuirl, J. M., and Sarter, N. B. 2006. Supporting trust calibration and the effective use of decision aids by presenting dynamic system confidence information. *Human Factors: The Journal of the Human Factors and Ergonomics Society* 48: 656–665.

Mercado, J., Rupp, M., Chen, J., et al. 2016. Intelligent agent transparency in human-agent teaming for multi-UxV management. *Human Factors: The Journal of the Human Factors and Ergonomics Society* 58: 401–415.

Meyer, J., & Lee, J. D. (2013). Trust, reliance, and compliance. In J. D. Lee & A. Kirlik (Eds.), *Oxford library of psychology. The Oxford handbook of cognitive engineering* (p. 109–124). Oxford University Press.

Miller, T. 2019. Explanation in artificial intelligence: Insights from the social sciences. *Artificial Intelligence* 267: 1–38.

Nam, C., Walker, P., Li, H., et al. 2020. Models of trust in human control of swarms with varied levels of autonomy. *IEEE Transactions on Human-Machine Systems* 50: 194–204.

Pynadath, D. V., Barnes, M. J., Wang, N., et al. 2018. Transparency communication for machine learning in human-automation interaction. In *Human and Machine Learning: Visible, Explainable, Trustworthy and Transparent*, eds. J. Zhou and F. Chen. New York, NY: Springer.

Rao, A. S., and Georgeff, M. P. 1995. *BDI agents: From theory to practice. Proceedings of the First International Conference on Multiagent Systems* (pp. 312–319). Palo Alto, CA: Association for the Advancement of Artificial Intelligence Press.

Roth, G., Schulte, A., Schmitt, F., et al. 2020. Transparency for a workload-adaptive cognitive agent in a manned–unmanned teaming application. *IEEE Transactions on Human-Machine Systems* 50: 225–233.

Roundtree, K. A., Goodrich, M. A., and Adams, J. A. 2019. Transparency: Transitioning from human–machine systems to human-swarm systems. *Journal of Cognitive Engineering and Decision Making* 13: 171–195.

Roundtree, K. A., Cody, J. R., Leaf, J., et al. 2020. Human-collective visualization transparency (preprint https://arxiv.org/abs/2003.10681)

Selkowitz, A., Lakhmani, S., and Chen, J. Y. C. 2017. Using agent transparency to support situation awareness of the Autonomous Squad Member. *Cognitive Systems Research* 46: 13–25.

Stowers, K., Kasdaglis, N., Rupp, M. A., et al. 2020. The IMPACT of agent transparency on human performance. *IEEE Transactions on Human-Machine Systems* 50: 245–253.

Vered, M., Howe, P., Miller, T., et al. 2020. Demand-driven transparency for monitoring intelligent agents. *IEEE Transactions on Human-Machine Systems* 50: 264–275.

Wang, N., Pynadath, D., and Hill, S. 2016. *Trust calibration within a human–robot team: Comparing automatically generated explanations.* In *Proceedings of the 11th ACM/ IEEE International Conference on Human-Robot Interaction* (pp. 109–116). Piscataway, NJ: IEEE Press.

Wang, N., Pynadath, D. V., Hill, S. G., et al. 2017. The dynamics of human-agent trust with POMDP-generated explanations. *In Proceedings of the International Conference on Intelligent Virtual Agents* (pp. 459–462).

Wohleber, R., Stowers, K., Chen, J. Y. C., et al. 2017. *Effects of agent transparency and communication framing on human-agent teaming. Proceedings of IEEE Conference on Systems, Man, & Cybernetics* (pp. 3427–3432), Banff, Canada (5–8 Oct, 2017).

Wright, J., Chen, J. Y. C., and Lakhmani, S. 2020. Agent transparency and reliability in human-robot interaction: The influence on user confidence and perceived reliability. *IEEE Transactions on Human-Machine Systems* 50: 254–263.

Wynne, K. T., and Lyons, J. B. 2018. An integrative model of autonomous teammate-likeness. *Theoretical Issues in Ergonomics Science* 19: 353–374.

Zhou, J., & Chen, F. (2015). Making machine learning useable. *International Journal of Intelligent Systems Technologies and Applications*, 14(2): 91–109.

Zhou, J., Khawaja, M. A., Li, Z., et al. 2016. Making machine learning useable by revealing internal states update – a transparent approach. *International Journal on Computational Science and Engineering* 13: 378–389.

5 Smart Telehealth Systems for the Aging Population

Carl Markert, Jukrin Moon, and Farzan Sasangohar

CONTENTS

5.1 INTRODUCTION

The population of seniors (aged 65 or over) in the United States in 2014 numbered approximately 46 million people (15% of the total population) and is expected to grow to approximately 82 million people (22% of the total population) by 2040 (Colby and Ortman 2017). Ninety percent of these seniors desire to live independently or remain in their homes as they age (Khalfani-Cox 2017). *Aging in place* is defined as "the ability to live in one's own home and community safely, independently, and comfortably, regardless of age, income, or ability level" (National Center for Environmental Health 2017). Aging in place allows seniors to maintain contact with friends, relatives, and their community and can reduce costs for health care and assisted living care (Satariano, Scharlach, and Lindeman 2014).

Chronic diseases are long-term health conditions that require ongoing medical attention and/or limit one's activities of daily living (ADLs) (Centers for Disease Control and Prevention 2018). Chronic diseases are common among older adults

DOI: 10.1201/9781003215349-5

and were the leading cause of death for seniors in the United States in 2017 (Kochanek et al. 2019). They also create a substantial cost for seniors and the healthcare system; in 2010 healthcare costs for seniors were over 2–1/2 times the national average and accounted for over one third of healthcare costs in the United States (De Nardi et al. 2015).

The combination of the projected growth of the aging population, the desire of seniors to age in place, and the prevalence and cost of chronic diseases has led the U.S. federal government to embark on a national program to develop strategies and tools to support aging adults (National Science & Technology Council 2019). Access to health care is one of the six primary functional capabilities identified in the program as being critical to seniors who desire to live an independent lifestyle as they age.

Telehealth is a prevalent and growing healthcare delivery method which has shown promise in facilitating access to care and enabling independent-living seniors to have access to effective health care. In addition, telehealth is identified as a means of enabling self-management of one's health care by helping people manage their health, educating them on their health conditions, and helping them take a more active role in the monitoring and treatment of their health conditions. Tuckson et al. (2017) have attributed the rapid growth of telehealth to five key trends: recent growth in consumer technology market and sensors, advancements in electronic health records and clinical decision support tools with potential for telehealth integration, projected shortages in healthcare professional workforce, recent incentives to lower cost of care, and growth of consumerism (preferences for real-time and convenient access to care). The need for quarantine and social distancing associated with the coronavirus pandemic (COVID-19) of 2020 also helped frame the benefits of telehealth as medical practices, public health organizations, and other health-related institutions rapidly accelerated telehealth adoption out of necessity. A population which can particularly benefit from telehealth are seniors who desire to age in place while maintaining convenient access to healthcare resources. This chapter provides a review of telehealth systems for the aging population and the incorporation of artificial intelligence into such systems. The chapter will also cover the following:

- an overview of telehealth as it relates to various health conditions or diseases,
- evidence for effectiveness of telehealth for managing chronic conditions,
- an integrated telehealth systems framework which shows the interactions between human, technological, and process-related components from a systems perspective, and
- several areas of proposed future work related to advanced telehealth systems which are capable of incorporating artificial intelligence and machine learning to recognize trends, diagnose conditions, recommend interventions, and promote behavioral change.

Telehealth is an all-inclusive term for remote health care including both clinical and non-clinical healthcare services. The Center for Connected Health Policy defines telehealth as "a collection of means or methods for enhancing health care, public health and health education delivery and support using telecommunications technologies" (Center for Connected Health Policy 2019). *Telemedicine* is the use of telehealth technology to provide clinical services (e.g., medical therapy) to patients remotely (Hall 2012).

Telehealth systems may include (1) *remote patient monitoring (RPM)*, which is the use of telehealth technology for remotely monitoring a patient's physiological parameters (such as blood pressure, heart rate, and body weight) and biological parameters (such as blood sugar level) and transmitting patient physiological data to a remote monitoring (RM) facility (Bratan and Clarke 2005); (2) *remote activity monitoring (RAM)*, which is the use of monitoring technology to remotely monitor a person's mobility and ADLs such as steps taken and sleep patterns (Mahoney 2010); (3) *decision support systems* (DSSs), which are smart electronic systems that evaluate health-related data and make clinical and/or behavioral recommendations; and (4) *health coaching systems* (HCSs), which are manual or electronic systems that employ a patient-centered process to promote behavior change; they include goal-setting, education, encouragement, and feedback on health-related behaviors (Oliveira et al. 2017). Advanced telehealth systems are capable of incorporating artificial intelligence and machine learning to recognize trends, diagnose conditions, recommend interventions, and promote healthy habits.

In what follows we describe these systems in more detail, present an integrated model that demonstrates their interconnections, and review the documented evidence supporting their effectiveness.

5.2 REMOTE MONITORING

RM is an emerging technology that will play a big role in enabling seniors to age in place. RM uses electronic devices, sensors, and telecommunication technology to monitor a patient remotely instead of requiring the patient to visit a healthcare facility (Malasinghe, Ramzan, and Dahal 2019). The ability to routinely and continuously monitor a patient via RM can assist in the early detection of illnesses or the detection of deteriorating conditions. This in turn can prevent a condition from worsening or the need to hospitalize the patient, thus improving quality of life (QOL) and reducing the cost of health care.

RM systems consist of one or more of the following components: (1) RPM for monitoring health-related parameters such as vital signs; (2) remote activity monitoring (RAM) for monitoring ADLs; (3) RM of medication adherence; (4) electronic diaries or queries documenting a patient's symptoms, health condition and wellness status, and (5) smart homes. Figure 5.1 is an example of bluetooth-enabled equipment for monitoring a patient's blood pressure, heart rate and body weight remotely using a mobile app to transmit the monitored parameters to a RM facility.

5.2.1 REMOTE PATIENT MONITORING

Remote patient monitoring (RPM) uses electronic devices, sensors, and telecommunication technology to monitor a patient's health-related data (Malasinghe, Ramzan, and Dahal 2019). It usually involves bluetooth-enabled or internet-connected devices that automatically transmit monitored parameters to a RM facility. Physiological

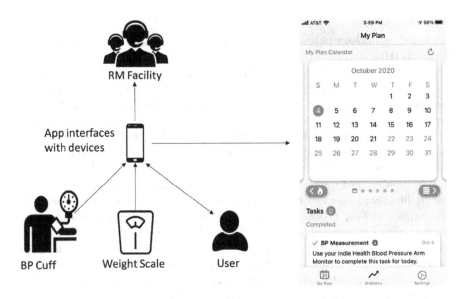

FIGURE 5.1 A representative remote monitoring system for hypertension consisting of Bluetooth-enabled equipment, integration with a mobile health application.

parameters monitored via RPM can provide healthcare professionals with an overview of the patient's health status or real-time values, including blood pressure, heart rate, body weight, body temperature, respiration rate, and pulse oximetry (Majumder et al. 2017). RPM has been used to monitor chronic conditions such as diabetes (Welch, Balder, and Zagarins 2015), hypertension (Logan et al. 2007), and mental health (McDonald et al. 2019).

The medical monitoring capabilities of RPM is especially well suited for seniors, a large majority of whom prefer to remain in their homes as they age (Khalfani-Cox 2017). Some advantages of RPM may include real-time monitoring of patients, continuous monitoring, early detection of illnesses, alerting features for patient and providers, prevention of worsening conditions, reductions in emergency care and hospitalizations, improved efficiency of health care, and potential cost savings (Malasinghe, Ramzan, and Dahal 2019).

5.2.2 REMOTE ACTIVITY MONITORING

Remote activity monitoring (RAM) uses sensors and wearable or mobile equipment to monitor information related to a patient's ADLs (Mahoney 2010), including mobility, eating, bathing, dressing, and toileting (National Science & Technology Council 2019). Mobility is an important ADL for older adults, as changes in mobility or activity are an indication of deteriorating health and wellness (National Science & Technology Council 2019), and RAM provides a means of monitoring a senior's mobility. RAM has been utilized for fall detection systems using sensors worn by a patient or installed in their homes (Malasinghe, Ramzan, and Dahal 2019). RAM can also use smart electricity meters combined with machine learning to monitor a person's electricity usage (Chalmers et al. 2016). Changes in usage could indicate behavioral changes in a person's routine indicative of a health-related issue. For example, changes in normal energy consumption at night might indicate a sleep disorder while changes in normal energy usage at mealtime might indicate an eating disorder.

5.2.3 RM FOR MEDICATION ADHERENCE

Seniors who take multiple or certain kinds of medications are at a higher risk of adverse drug reactions that could limit their ability to safely age in place (National Science & Technology Council 2019). Remote medication monitoring using the readily available consumer products such as electronic pill boxes provides the opportunity for real-time monitoring, intervention, and adherence feedback (National Science & Technology Council 2019). Research is ongoing into smart pill boxes or automatic pill reminders and dispensing systems that distribute and track medication dosage. One such pill box consists of various compartments to hold the pills, a timer or reminder system to alert the patient when it is time to take their medications, and a sensor to detect when the medicine has been removed from its compartment (Jabeena et al. 2017). When medicine is removed, the system registers that the patient has successfully taken the medicine. If the patient does not take their medication by the prescribed time, an alarm is initiated, and after a prescribed time period a reminder message is sent to them. If the patient still fails to take the medication after the reminder, a message is sent to their family members. This system tracks regular medication adherence, reduces the memory requirement to keep track of when medications are due, and helps ensure the proper medication and dosages are taken at the right time. Several similar technologies have been developed (such as the Sagely Smart XL Weekly Pill Organizer, Tricella Pillbox and Med-Q Automatic Pill Dispenser), and at the time of this writing the market for smart and internet-of-things (IoT)-enabled home medication management products is expanding rapidly.

5.2.4 ELECTRONIC DIARIES AND QUERIES

Electronic communication tools such as CareSense are increasingly used by healthcare systems to gather information on a patient's symptoms, health condition, and wellness status. Questions can be sent to a monitored person daily via a computer or mobile device for their response. Questions could be generic in nature or specific to a person's health condition (e.g., relating to heart failure symptoms for patients with chronic heart failure (Agboola et al. 2015) or to shortness of breath for patients with chronic obstructive pulmonary disorder (Dang, Dimmick, and Kelkar 2009). Electronic diaries can also be used to self-report symptoms and overall health status periodically (Pavel et al. 2015) for subsequent review and analysis by a healthcare provider.

5.2.5 SMART HOMES

Smart homes combine RPM, RAM, medication monitoring, and electronic diaries within living spaces designed with RM sensors and cameras connected to the IoT to enable aging in place (National Science & Technology Council 2019). For example, today's smart home technology includes installed sensors, actuators, and telecommunication equipment capable of monitoring a patient's ADLs with or without the use of cameras (Majumder et al. 2017; Chan et al. 2009). While smart home technologies raise privacy issues, a study by the Gerontological Society of America found that many seniors are willing to trade privacy for the ability to live an autonomous, independent lifestyle (Berridge 2016).

5.2.6 Chronic, Acute, and Ambulatory Motivations for RM

RM, of any of the varieties above, can be categorized based on the purpose for the monitoring. Chronic disease management involves the routine monitoring of patients suffering from one or more chronic diseases such as cardiovascular system diseases, respiratory system diseases, or diabetes (Bratan and Clarke 2005). The purpose of chronic disease management monitoring is for early detection of deteriorating conditions indicating the possible need for intervention.

Acute monitoring involves the near-continuous monitoring of patients with acute health conditions that could rapidly deteriorate (Bratan and Clarke 2005). This could be for a serious illness, severe infections, or following surgical procedures. Monitored data need to be sent on a near-continuous basis to healthcare workers, with provision for rapid clinical intervention if required. Acute monitoring can be used in a home setting but is probably best utilized in a healthcare setting or assisted living facility.

Ambulatory monitoring entails vital sign monitoring in an otherwise healthy person (Bratan and Clarke 2005). The purpose of ambulatory monitoring is to identify a sudden change in a person's health status that might be an early indicator of an impending serious health condition. Wellness and safety monitoring (or lifestyle monitoring) includes ambulatory monitoring plus ADL monitoring via RAM sensors (Hargreaves et al. 2017). The purpose of wellness and safety monitoring is to identify a change in a person's health and wellness that might require intervention before their next routine healthcare checkup.

5.3 DECISION SUPPORT SYSTEMS

Unlike the early telehealth systems that relied on telecommunication data between patients and their healthcare providers, today's smart telehealth systems rely on telemonitoring data acquired through the mobile or smart home technologies combined with multi-modal or multi-parametric sensors (Colantonio et al. 2015). Given the large datasets collected via longitudinal RM, clinical professionals need decision-support tools to interpret the raw data and detect the abnormality in patients' health status with increased speed and capacity (Basilakis et al. 2010). As such, DSSs combined with RM are becoming some of the most prevalent and fastest growing telehealth interventions.

DSSs are computerized systems which evaluate data collected via RM and transform the data into useful information regarding the patients' health and wellness. Based on the clinical interpretation of large datasets collected via RM and self-reports, DSSs provide healthcare-related recommendations to healthcare providers,

HCSs, and/or the patient as appropriate. Datasets of the patients' physiological, behavioral, and medication status serve as inputs to the DSS. The DSS processes monitored data and compares it to standard clinical guidelines such as the American College of Cardiology/American Heart Association (ACC/AHA) guidelines for hypertension (Ioannidis 2018). Output from the DSS includes recommendations on healthcare interventions.

Transforming the raw sensor data to useful information (processing the inputs to produce the outputs of DSSs) is particularly challenging due to limited sensing quality. Sensing quality is limited because the remotely monitored data, using unobtrusive sensors, usually represent "only surrogate markers of the phenomenon of interest"; accordingly, the inference made from imperfect sensing becomes compromised (Pavel et al. 2015). As such, some efforts have combined information from multi-modal or multi-parametric sensors (Zhang, Ren, and Shi 2013) with context awareness (Noury et al. 2011).

DSSs vary depending on the level of automation in clinical data interpretation (Colantonio et al. 2015). There can be a simple manual DSS in which no automation is involved where clinical professionals interpret the telemonitored data and make an assessment of the patients' health status (Ding et al. 2012). On the other hand, there can be an artificial intelligence-augmented DSS in which data interpretation is augmented by partial or full automation (Basilakis et al. 2010; Song et al. 2010; van der Heijden et al. 2013).

A wide range of artificial intelligence-augmented DSSs have been proposed for the aging population, particularly with the technological advancement of wearable devices (Agoulmine et al. 2011; Deen 2015; Lin et al. 2010; Rashidi and Mihailidis 2013) and smart homes (Majumder et al. 2017). Since it is not always possible to directly sense an older person's ability to conduct ADLs, a DSS can infer that a person has made a meal from sensors associated with kitchen lighting, operation of the range or microwave, and opening/closing the refrigerator or pantry (Pavel et al. 2015). The DSS can also initiate an emergency notification to 911 if certain threshold values of monitored parameters are exceeded (Choudhury et al. 2015). Some other examples involve smart homes with automated activity and fall detection DSSs (Zhang, Ren, and Shi 2013) and automated sleep monitoring systems (Barsocchi et al. 2016).

The implications of an artificial intelligence-augmented DSS combined with RM are significant. Despite the remaining challenges associated with recording and processing reliable clinical variables in automated and unsupervised data collection environments (Basilakis et al. 2010), automation in data interpretation can reduce the workload of clinical professionals, and ultimately, the cost of health care. Finkelstein et al. (2013), for instance, demonstrated the potential benefits of automated DSSs to assist clinical transplant nurses in offloading the time-consuming and fatiguing surveillance process and assisting in triage decisions regarding clinical intervention for lung transplant patients participating in a RM program. This leads to cost savings because of the expanded bandwidth of the clinical workforce and the healthcare systems to monitor more patients when equipped with the automated DSSs (Finkelstein et al. 2013).

5.4   HEALTH COACHING SYSTEMS

Health coaching or wellness coaching is becoming a recognized strategy to help people improve their health and well-being. HCSs employ a patient-centered process to promote behavior change and include goal-setting, education, encouragement, and feedback on health-related behaviors (Oliveira et al. 2017). Health coaching is a collaborative effort between a trained health coach and a patient and includes: (1) education about a person's health condition; (2) an understanding of a person's strengths, challenges and motivation to change behaviors; (3) realistic goal-setting; and (4) feedback on adherence, progress and achievement of goals (Clark et al. 2014). Whereas disease management processes focus on the treatment of a specific disease, HCSs focus on the patient's health-related behaviors (Jonk et al. 2015). Forms of health coaching include periodic health tips, educational material, healthcare suggestions, and encouragement and feedback on achieving pre-established goals. HCSs may be manual (human health coach only), partially automated (human health coach augmented by a smart system such as a DSS) or fully automated (using smart systems combined with artificial intelligence and machine learning to generate coaching messages for the patient).

An example of a manual HCS is a home pulmonary rehabilitation (PR) program with health coaching (Benzo et al. 2018). The pilot coaching system described by Benzo et al. (2018) consists of a mobile device (tablet computer), an activity monitor, and a pulse oximeter plus the trained PR health coach. Patients performed exercises as demonstrated on the mobile device while wearing the pulse oximeter. After exercising, patients completed a wellness questionnaire on the mobile device to record their well-being, any breathing difficulties, and overall energy level. Patients also wore a pedometer continuously to record step counts. Motivation was provided to the patients through instantaneous feedback from the oximeter while exercising (heart rate and blood oxygen level) and cumulative steps from the activity monitor throughout the day. Data from the pulse oximeter, activity monitor, and daily wellness questionnaire were uploaded to a RM facility for the health coach to review. After reviewing the monitored data and daily questionnaires, the health coach called the patient weekly to discuss their results, adherence to the program and progress. The health coach used motivational interviewing techniques to promote behavioral health change. During this call, the health coach and the patient agreed on goals for the following week. The end goal of this RM plus health coaching process was a permanent behavioral change with long-term adherence to a healthier lifestyle that included daily physical activity.

Partially automated HCSs include human health coaches augmented by various levels of automation for data collection and analysis, while fully automated HCSs minimize the involvement of a human coach after initial program establishment and goal setting. An example of a fully automated HCS is a diabetes management program that includes activity monitoring and personalized messages to encourage physical activity based on a reinforcement learning (RL) algorithm (Yom-Tov et al. 2017). The coaching system described in Yom-Tov et al. consisted of a mobile app that monitored physical activity, a series of coaching messages and an RL algorithm to select a daily message. RL was used by the algorithm to select a daily coaching message based on the patient's demographics, past activity, expected activity, and previous messages. This coaching system did not require daily manual input from the patient or a health coach. The algorithm automatically determined an appropriate coaching message with the goal of increasing the patient's daily walking duration and distance.

Health coaching using robotics has been demonstrated to be a successful telehealth intervention method in older adults (Bakas et al. 2018). A feasibility study by Bakas et al. (2018) evaluated the use of a remote-controlled robot to provide communication between the health coach and the patient. A major benefit of the robot over the use of a computer or mobile device is elimination of the need for the older adult to adapt to computer technology. A nurse, trained in health coaching, interacted with the patient three times a week via the remote-controlled robot. This study (11 intervention group participants and 10 control group participants) showed medium to large effect sizes for improving self-management of chronic conditions, sleep quality, and QOL as well as a reduction in unhealthy days and depression symptoms in older adults. Overall, this study demonstrated the positive aspects of goal setting and action plan development and monitoring as part of the health coaching program.

HCSs plus RPM can also be used as a healthcare intervention following hospitalization for a major illness. The Better Effectiveness After Transition-Heart Failure (BEAT-HF) randomized clinical trial (Ong et al. 2016) combined daily RPM (blood pressure, heart rate, symptoms, and weight) with routine health coaching telephone calls from a registered nurse. The BEAT-HF study found that neither re-hospitalizations nor 180-day mortality were significantly affected by the intervention; however, the study noted that QOL may have been improved for the intervention group.

One of the challenges of any health coaching intervention is sustaining results after health coaching has ended. A 2016 prospective randomized controlled trial (Sharma et al. 2016) evaluated the sustainability of health coaching after participation in a 12-month health coaching program for patients with diabetes, hypertension, and/or hyperlipidemia (enrolling 224 patients in the intervention with 217 in a usual care control group). At the 24-month point, patients in the health coaching intervention group showed only minimal declines in clinical measurements with the exception of HbA1c for diabetic patients. The primary investigators of this study concluded that the effects of health coaching can be sustained for some conditions but may require continuing support for other conditions such as diabetes.

5.5   INTEGRATED TELEHEALTH SYSTEM

Advanced telehealth systems are capable of incorporating artificial intelligence and machine learning to recognize trends, diagnose conditions, recommend interventions, and promote behavioral change. Thus far, we have looked into the three most common interventions in advanced telehealth systems: RM, DSSs, and HCSs. While these telehealth systems provide unique benefits, several interconnected components provide an opportunity for an integrated telehealth system that draws from advantages of each. Figure 5.2 proposes a model for such integrated approach for providing telehealth services to seniors using all three intervention components. While not all components included in this model may exist in all telehealth systems, such a holistic model provides a generic representation that may inform future system design and implementation efforts.

In the integrated telehealth system, those three interventional components feed forward and back into each other. RM acquires patients' physiological, activity, and medication status data. The raw datasets collected via RM or self-reports feed forward into the manual or artificial intelligence-augmented DSS. The DSS interprets the data and produces clinically relevant inferences or predictions on patient health. The resulting information from the DSS may feed back to the patient, or feed forward into healthcare providers, family members or HCSs as appropriate. The information provided by DSSs becomes the basis for generating healthcare recommendations. The information also becomes the basis for making health coaching recommendations and ultimately changing the patients' health related behaviors. The changes in patients' behaviors influence the patients' physiological, behavioral, and medication status, which may feed back into RM and DSSs. As such, the connections among three interventional components provide a closed loop between data acquisition and intervention (Pavel et al. 2015)

A fully integrated telehealth system ensures a patient receives tailored coaching interventions to change or sustain behaviors and timely feedbacks from the computational assessment of remotely monitored status, activity, and context. Given the integration of the three telehealth intervention components, the patient, as well as healthcare providers and family members, can easily access the monitored data, the transformed information, healthcare recommendations, and HCS recommendations when desired. With such increased access to the system components, the patient in the integrated telehealth system can play a greater role in self-management of their health care and health behavior change. According to Pavel (2015), such an integrated

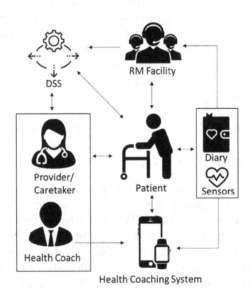

FIGURE 5.2   Integrated telehealth system (adapted from Pavel et al. 2015).

telehealth system would serve as an important tool for seniors to better learn and adhere to health behaviors which can enable them to age in place.

5.6 REVIEW METHODS

While various aforementioned telehealth systems for the aging populations have been utilized and discussed, one should pay careful attention to the available evidence suggesting their efficacy. After all, using technology just because it is viable might result in rejection and frustration. Therefore, we reviewed smart telehealth interventions for seniors, with a particular focus on examining if and to what extent they were evaluated to be effective. Initially, a scoping literature review was conducted to compile and synthesize knowledge related to smart telehealth interventions for seniors. The CINAHL and MEDLINE (Ovid) databases were searched using free-text search terms relevant to smart telehealth interventions for seniors (including terms such as monitoring, physiologic, remote sensing technology, RM, wearable, and aged/elder/senior). The initial search yielded about 700 articles published after 2008 that met the inclusion criteria.

After applying exclusion criteria (including duplicates, foreign language only, cardiac implanted electronic devices, not in-home health care, effectiveness not evaluated, and health conditions other than cardiovascular diseases, respiratory diseases, or diabetes) to the title/abstract screening and full-text assessments, 50 records were selected for the final review. Each of these records were coded as one of the three telehealth systems discussed in this chapter, i.e., RM, DSSs, and HCSs. Additional records related to smart telehealth systems for the aging population were retrieved and reviewed to augment the initial scoping review and are included as references in this chapter. The findings of this review are partly published in Markert, Rao, and Sasangohar (2019). Here we provide a summary of findings based on a 2016 U.S. Department of Health and Human Services report that provided an overview of the current state of telehealth in the United States (Totten et al. 2016). The report presented an abbreviated review of 58 systematic reviews evaluating the effectiveness of telehealth on clinical outcomes, healthcare utilization, or costs.

5.7 EFFECTIVENESS OF TELEHEALTH TO SUPPORT SUCCESSFUL AGING IN PLACE

One of the challenges of broadly implementing telehealth in the senior population is demonstrating its effectiveness. There are various methods available to assess effectiveness of telehealth for seniors. Some of the more common metrics are clinical effectiveness, QOL effectiveness, cost effectiveness, and successful transition to self-management of one's health care.

Clinical effectiveness is the ability of a healthcare intervention method to improve or maintain a person's health or wellness status. Common metrics to assess clinical effectiveness include (1) management of the illness/symptoms, (2) number of emergency room visits, (3) number of hospital admissions/readmissions, (4) length of stay in a hospital, (5) mortality rate or time between a serious illness and death, and (6) stamina or exercise tolerance.

QOL is a state of mind or self-reported perception of health status and not merely characterized by the absence of diseases (Moriarty, Zack, and Kobau 2003). Some of methods available to measure QOL are (1) the Rand short form 36 (SF-36) questionnaire (Hays, Sherbourne, and Mazel 1993), (2) Minnesota living with heart failure questionnaire (Garin et al. 2013), (3) COPD assessment test (Dodd et al. 2012), and (4) subjective assessments.

Cost effectiveness is a measure of the benefit received (clinical outcome or QOL) from a healthcare intervention method versus the cost of the intervention, in particular, the monetary cost. Components of telehealth cost include (1) telehealth equipment costs, (2) monitoring costs, (3) direct healthcare costs, (4) healthcare facility use costs, (5) costs to Medicare, and (6) medication costs.

The successful transition to self-management of one's health care can be assessed by the long-term sustainability or adherence to a health and wellness program upon termination of the formal monitoring and coaching aspects of the program. Self-management of one's health is the process of learning skills and knowledge required to manage one's symptoms and to reduce risk factors associated with one's health condition (Whittemore and Dixon 2008). The skills and knowledge learned need to be converted into healthy behaviors and lifestyle changes to solidify those behaviors indefinitely. A formal telehealth system that includes monitoring and coaching can provide the skills and knowledge needed to effect behavior change, and effectiveness is an assessment of the sustainability and adherence to the change without the coaching aspect of an integrated telehealth system.

According to Totten et al. (2016), telehealth is effective for the following situations: (1) RPM for patients with chronic conditions, (2) health coaching (communicating with and counseling) for patients with chronic conditions, and (3) psychotherapy for patients with mental health problems. The report found that patients with chronic diseases such as cardiovascular and respiratory diseases specifically benefitted from RPM and health coaching as measured by mortality, QOL, and hospital admissions. One of the areas identified for future research was the integration of behavioral and physical health.

According to the same report, older patients with chronic diseases are a frequently targeted population for telehealth because this population usually requires more frequent healthcare visits and clinical support to self-manage their chronic conditions (Totten et al. 2016). Telehealth for seniors can help overcome barriers to healthcare services, prevent worsening of conditions, potentially avoid emergency room visits, and reduce healthcare costs while increasing a patient's wellness and QOL. The final conclusion of the report was that telehealth was generally effective in the area of clinical outcomes for patients with chronic conditions. Fewer studies included in the review focused on the economic benefit of telehealth, with approximately half of the applicable studies showing some healthcare utilization and cost benefit while the others were inconclusive or found no benefit. The review concluded that additional studies are needed on the healthcare utilization and cost–benefit balance of telehealth.

We also reviewed the effectiveness of health coaching combined with telehealth for older adults (Markert, Rao, and Sasangohar 2019). Clinical effectiveness was satisfactorily reported in majority of the studies, and QOL effectiveness was satisfactorily reported in most of studies that evaluated QOL. Cost effectiveness was only evaluated in two studies and found to be ineffective in both studies. Major gaps identified in the review included the need for additional research in the following areas: (1) cost effectiveness of telehealth (concurring with Totten et al. 2016), (2) long-term sustainability of behavior changes after formal telehealth program termination, and (3) inclusion of depression monitoring and intervention as part of telehealth programs.

5.8 CONCLUSIONS

Telehealth systems provide tangible benefits for the rapidly growing aging population in the United States and beyond. Given the forced adoption of telehealth by healthcare systems during the COVID-19 pandemic, a wide-scale mental model of such systems has formed and benefits have become more tangible. Therefore, the high utilization rate for these technologies is expected to remain beyond the pandemic. In this chapter, we reviewed three main categories of telehealth systems for the aging population and proposed an integrative model that illustrated the future of remote data acquisition, patient–provider interaction, and clinical inference. While several components of this model are already in place, the integration between such potentially complementary systems may provide an opportunity to provide an infrastructure with sustainable and tangible benefits for the aging population.

Several areas of future work are proposed related to advanced telehealth systems which are capable of incorporating artificial intelligence and machine learning to recognize trends, diagnose conditions, recommend interventions, and promote behavioral change. Telehealth programs have been shown to have clinical and QOL benefits while people are actively enrolled in the program; however, those benefits might not be sustained after disenrollment from a formal program. Additional post-termination studies are needed to assess the long-term sustainability of the positive effects brought about by a telehealth program. This research can facilitate a program designed to help transition seniors from telehealth to self-management of their health and wellness. In addition, while telehealth programs have been shown to contribute to an improved QOL, insufficient evidence exists to conclude if the improved QOL is due to the psychological benefit of being monitored or feedback provided by the telehealth system. Additional studies are needed to evaluate the psychological benefit of telehealth on QOL for seniors and whether there is a relationship between the psychological benefit and one's overall health and wellness.

With technological advances in RPM and RAM sensors and big data analytics, the capabilities of DSSs to analyze data and identify interventions when necessary will be enhanced. Additional research should be conducted on DSSs as part of the integrated healthcare solution for seniors. Additional research is also needed to assess the capability of automated HCSs versus human coaches to affect health behavior changes. In addition, cost effectiveness of automated health coaching needs to be assessed against human-only health coaching methods. The results of this research would inform the future direction of automated health coaching. Finally, cost effectiveness of telehealth programs has not been adequately demonstrated. Additional cost studies are necessary to provide a solid basis for implementing telehealth on a larger scale.

REFERENCES

Agboola, Stephen, Kamal Jethwani, Kholoud Khateeb, Stephanie Moore, and Joseph Kvedar. 2015. "Heart Failure Remote Monitoring: Evidence from the Retrospective Evaluation of a Real-World Remote Monitoring Program." *Journal of Medical Internet Research* 17(4). doi:10.2196/jmir.4417.

Agoulmine, Nazim, M. Jamal Deen, Jeong-Soo Lee, and M. Meyyappan. 2011. "U-Health Smart Home." *IEEE Nanotechnology Magazine* 5(3): 6–11. doi:10.1109/MNANO .2011.941951.

Bakas, Tamilyn, Debi Sampsel, Jahmeel Israel, Ameya Chamnikar, Barbara Bodnarik, John Greer Clark, Megan Gresham Ulrich, and Dieter Vanderelst. 2018. "Using Telehealth to Optimize Healthy Independent Living for Older Adults: A Feasibility Study." *Geriatric Nursing* 39(5): 566–573.

Barsocchi, Paolo, Monica Bianchini, Antonino Crivello, Davide La Rosa, Filippo Palumbo, and Franco Scarselli. 2016. *"An Unobtrusive Sleep Monitoring System for the Human Sleep Behaviour Understanding."* In *2016 7th IEEE International Conference on Cognitive Infocommunications (CogInfoCom)*, 000091–000096. doi:10.1109/ CogInfoCom.2016.7804531.

Basilakis, Jim, Nigel H. Lovell, Stephen J. Redmond, and Branko G. Celler. 2010. "Design of a Decision-Support Architecture for Management of Remotely Monitored Patients." *IEEE Transactions on Information Technology in Biomedicine* 14(5): 1216–1226. doi:10.1109/TITB.2010.2055881.

Benzo, Roberto P., Kevin M. Kramer, Johanna P. Hoult, Paige M. Anderson, Ivonne M. Begue, and Sara J. Seifert. 2018. "Development and Feasibility of a Home Pulmonary Rehabilitation Program with Health Coaching." *Respiratory Care* 63(2): 131–140. doi:10.4187/respcare.05690.

Berridge, Clara. 2016. "Breathing Room in Monitored Space: The Impact of Passive Monitoring Technology on Privacy in Independent Living." *Gerontologist* 56(5): 807–816.

Bratan, T., and M. Clarke. 2005. *"Towards the Design of a Generic Systems Architecture for Remote Patient Monitoring."* In *2005 IEEE Engineering in Medicine and Biology 27th Annual Conference*, 106–109. Shanghai, China: IEEE. doi:10.1109/IEMBS .2005.1616353.

Center for Connected Health Policy. 2019. "About Telehealth." https://www.cchpca.org/about/ about-telehealth.

Centers for Disease Control and Prevention. 2018. "About Chronic Disease." September 5. https://www.cdc.gov/chronicdisease/about/index.htm.

Chalmers, Carl, William Hurst, Michael Mackay, and Paul Fergus. 2016. *"Smart Monitoring: An Intelligent System to Facilitate Health Care across an Ageing Population."* In *EMERGING 2016: The Eighth International Conference on Emerging Networks and Systems Intelligence*, 34–39. IARIA XPS Press.

Chan, Marie, Eric Campo, Daniel Estève, and Jean-Yves Fourniols. 2009. "Smart Homes - Current Features and Future Perspectives." *Maturitas* 64(2): 90–97.

Choudhury, Biplav, Tameem S. Choudhury, Aniket Pramanik, Wasim Arif, and J. Mehedi. 2015. *"Design and Implementation of an SMS Based Home Security System."* In *2015 IEEE International Conference on Electrical, Computer and Communication Technologies (ICECCT)*, 1–7. Coimbatore, India: IEEE. doi:10.1109/ICECCT.2015.7226115.

Clark, Matthew M., Karleah L. Bradley, Sarah M. Jenkins, Emily A. Mettler, Brent G. Larson, Heather R. Preston, Juliette T. Liesinger, Brooke L. Werneburg, Philip T. Hagen, and Ann M. Harris. 2014. "The Effectiveness of Wellness Coaching for Improving Quality of Life." *Mayo Clinic Proceedings* 89: 1537–1544.

Colantonio, S., L. Govoni, R. L. Dellacà, M. Martinelli, O. Salvetti, and M. Vitacca. 2015. "Decision Making Concepts for the Remote, Personalized Evaluation of COPD Patients' Health Status." *Methods of Information in Medicine* 54(3): 240–247. doi:10.3414/ME13-02-0038.

Colby, Sandra L., and Jennifer M. Ortman. 2017. "Projections of the Size and Composition of the US Population: 2014 to 2060." U.S. Census Bureau. http://wedocs.unep.org/bitstream/handle/20.500.11822/20152/colby_population.pdf.

Dang, Stuti, Susan Dimmick, and Geetanjali Kelkar. 2009. "Evaluating the Evidence Base for the Use of Home Telehealth Remote Monitoring in Elderly with Heart Failure." *Telemedicine and E-Health* 15(8): 783–796. doi:10.1089/tmj.2009.0028.

De Nardi, Mariacristina, Eric French, John Bailey Jones, and Jeremy McCauley. 2015. "Medical Spending of the U.S. Elderly." Working Paper 21270. National Bureau of Economic Research. doi:10.3386/w21270.

Deen, M. Jamal. 2015. "Information and Communications Technologies for Elderly Ubiquitous Healthcare in a Smart Home." *Personal and Ubiquitous Computing* 19(3): 573–599. doi:10.1007/s00779-015-0856-x.

Ding, H., Y. Moodley, Y. Kanagasingam, and M. Karunanithi. 2012. "*A Mobile-Health System to Manage Chronic Obstructive Pulmonary Disease Patients at Home.*" In *2012 Annual International Conference of the IEEE Engineering in Medicine and Biology Society*, 2178–2181. doi:10.1109/EMBC.2012.6346393.

Dodd, James W., Phillippa L. Marns, Amy L. Clark, Karen A. Ingram, Ria P Fowler, Jane L. Canavan, Mehul S. Patel, Samantha S. C. Kon, Nicholas S. Hopkinson, and Michael I. Polkey. 2012. "The COPD Assessment Test (CAT): Short-and Medium-Term Response to Pulmonary Rehabilitation." *COPD: Journal of Chronic Obstructive Pulmonary Disease* 9(4): 390–394.

Finkelstein, Stanley M., Bruce R. Lindgren, William Robiner, Ruth Lindquist, Marshall Hertz, Bradley P. Carlin, and Arin VanWormer. 2013. "A Randomized Controlled Trial Comparing Health and Quality of Life of Lung Transplant Recipients Following Nurse and Computer-Based Triage Utilizing Home Spirometry Monitoring." *Telemedicine Journal and E-Health* 19(12): 897–903.

Garin, Olatz, Montse Ferrer, Àngels Pont, Ingela Wiklund, Eric Van Ganse, Gemma Vilagut, Josué Almansa, Aida Ribera, and Jordi Alonso. 2013. "Evidence on the Global Measurement Model of the Minnesota Living with Heart Failure Questionnaire." *Quality of Life Research* 22(10): 2675–2684.

Hall, Amanda K. 2012. "Healthy Aging 2.0: The Potential of New Media and Technology." *Preventing Chronic Disease* 9. doi:10.5888/pcd9.110241.

Hargreaves, Sarah, Mark S. Hawley, Annette Haywood, and Pamela M. Enderby. 2017. "Informing the Design of 'Lifestyle Monitoring' Technology for the Detection of Health Deterioration in Long-Term Conditions: A Qualitative Study of People Living with Heart Failure." *Journal of Medical Internet Research* 19(6): e231. doi:10.2196/jmir.6931.

Hays, Ron D., Cathy Donald Sherbourne, and Rebecca M. Mazel. 1993. "The Rand 36-item Health Survey 1.0." *Health Economics* 2(3): 217–227.

van der Heijden, Maarten, Peter J. F. Lucas, Bas Lijnse, Yvonne F. Heijdra, and Tjard R. J. Schermer. 2013. "An Autonomous Mobile System for the Management of COPD." *Journal of Biomedical Informatics* 46(3): 458–469. doi:10.1016/j.jbi.2013.03.003.

Ioannidis, John PA. 2018. "Diagnosis and Treatment of Hypertension in the 2017 ACC/AHA Guidelines and in the Real World." *JAMA* 319(2): 115–116.

Jabeena, Afthab, Animesh Kumar Sahu, Rohit Roy, and N. Sardar Basha. 2017. "*Automatic Pill Reminder for Easy Supervision.*" In *2017 International Conference on Intelligent Sustainable Systems (ICISS)*, 630–637. IEEE.

Jonk, Yvonne, Karen Lawson, Heidi O'Connor, Kirsten S. Riise, David Eisenberg, Bryan Dowd, and Mary J. Kreitzer. 2015. "How Effective Is Health Coaching in Reducing Health Services Expenditures?" *Medical Care* 53(2): 133–140. doi:10.1097/MLR.0000000000000287.

Khalfani-Cox, L. 2017. "What Are the Costs of Aging in Place?" AARP. https://www.aarp.org/money/budgeting-saving/info-2017/costs-of-aging-in-place.html.

Kochanek, Kenneth D, Sherry L Murphy, Jiaquan Xu, and Elizabeth Arias. 2019. "Deaths: Final Data for 2017." *National Vital Statistics Reports*; vol 68 no 9. Hyattsville, MD: National Center for Health Statistics. https://www.cdc.gov/nchs/data/nvsr/nvsr68/nvsr68_09-508.pdf.

Lin, Chin-Teng, Kuan-Cheng Chang, Chun-Ling Lin, Chia-Cheng Chiang, Shao-Wei Lu, Shih-Sheng Chang, Bor-Shyh Lin, et al. 2010. "An Intelligent Telecardiology System Using a Wearable and Wireless ECG to Detect Atrial Fibrillation." *IEEE Transactions on Information Technology in Biomedicine* 14(3): 726–733. doi:10.1109/TITB.2010.2047401.

Logan, Alexander G, Warren J McIsaac, Andras Tisler, M Jane Irvine, Allison Saunders, Andrea Dunai, Carlos A Rizo, Denice S Feig, Melinda Hamill, and Mathieu Trudel. 2007. "Mobile Phone–Based Remote Patient Monitoring System for Management of Hypertension in Diabetic Patients." *American Journal of Hypertension* 20(9): 942–948.

Mahoney, Diane Feeney. 2010. "An Evidence-Based Adoption of Technology Model for Remote Monitoring of Elders' Daily Activities." *Ageing International* 36(1): 66–81. doi:10.1007/s12126-010-9073-0.

Majumder, Sumit, Emad Aghayi, Moein Noferesti, Hamidreza Memarzadeh-Tehran, Tapas Mondal, Zhibo Pang, and M Jamal Deen. 2017. "Smart Homes for Elderly Healthcare - Recent Advances and Research Challenges." *Sensors (Basel, Switzerland)* 17(11): 2496. doi:10.3390/s17112496.

Malasinghe, Lakmini P., Naeem Ramzan, and Keshav Dahal. 2019. "Remote Patient Monitoring: A Comprehensive Study." *Journal of Ambient Intelligence and Humanized Computing* 10(1): 57–76.

Markert, Carl, Arjun H. Rao, and Farzan Sasangohar. 2019. "Combining Health Coaching with Telehealth: A Scoping Review." *Proceedings of the Human Factors and Ergonomics Society Annual Meeting* 63(1): 16–16. doi:10.1177/1071181319631266.

McDonald, Anthony D, Farzan Sasangohar, Ashish Jatav, and Arjun H Rao. 2019. "Continuous Monitoring and Detection of Post-Traumatic Stress Disorder (PTSD) Triggers among Veterans: A Supervised Machine Learning Approach." *IISE Transactions on Healthcare Systems Engineering* 9(3): 201–211.

Moriarty, David G., Mathew M. Zack, and Rosemarie Kobau. 2003. "The Centers for Disease Control and Prevention's Healthy Days Measures – Population Tracking of Perceived Physical and Mental Health over Time." *Health and Quality of Life Outcomes* 1(1): 37. doi:10.1186/1477-7525-1-37.

National Center for Environmental Health. 2017. "CDC - Healthy Places - Healthy Places Terminology." December 11. https://www.cdc.gov/healthyplaces/terminology.htm.

National Science & Technology Council. 2019. "*Emerging Technologies to Support an Aging Population*." Washington, DC: Author. https://www.whitehouse.gov/wp-content/uploads/2019/03/Emerging-Tech-to-Support-Aging-2019.pdf.

Noury, Norbert, Marc Berenguer, Henri Teyssier, Marie-Jeanne Bouzid, and Michel Giordani. 2011. "Building an Index of Activity of Inhabitants From Their Activity on the Residential Electrical Power Line." *IEEE Transactions on Information Technology in Biomedicine* 15(5): 758–766. doi:10.1109/TITB.2011.2138149.

Oliveira, Juliana S., Catherine Sherrington, Anita B. Amorim, Amabile B. Dario, and Anne Tiedemann. 2017. "What Is the Effect of Health Coaching on Physical Activity Participation in People Aged 60 Years and over? A Systematic Review of Randomised Controlled Trials." *British Journal of Sports Medicine* 51(19): 1425–1432. doi:10.1136/bjsports-2016-096943.

Ong, Michael K., Patrick S. Romano, Sarah Edgington, Harriet U. Aronow, Andrew D. Auerbach, Jeanne T Black, Teresa De Marco, Jose J Escarce, Lorraine S Evangelista, and Barbara Hanna. 2016. "Effectiveness of Remote Patient Monitoring after Discharge of Hospitalized Patients with Heart Failure: The Better Effectiveness After Transition – Heart Failure (BEAT-HF) Randomized Clinical Trial." *JAMA Internal Medicine* [Erratum appears in JAMA Intern Med. 2016, 176(4): 568; PMID: 26974880], 176(3): 310–318. doi:10.1001/jamainternmed.2015.7712.

Pavel, Misha, Holly B Jimison, Ilkka Korhonen, Christine M Gordon, and Niilo Saranummi. 2015. "Behavioral Informatics and Computational Modeling in Support of Proactive Health Management and Care." *IEEE Transactions on Bio-Medical Engineering* 62(12): 2763–2775. doi:10.1109/TBME.2015.2484286.

Rashidi, Parisa, and Alex Mihailidis. 2013. "A Survey on Ambient-Assisted Living Tools for Older Adults." *IEEE Journal of Biomedical and Health Informatics* 17(3): 579–590. doi:10.1109/JBHI.2012.2234129.

Satariano, William A., Andrew E. Scharlach, and David Lindeman. 2014. "Aging, Place, and Technology: Toward Improving Access and Wellness in Older Populations." *Journal of Aging and Health* 26(8): 1373–1389. doi:10.1177/0898264314543470.

Sharma, Anjana E., Rachel Willard-Grace, Danielle Hessler, Thomas Bodenheimer, and David H. Thom. 2016. "What Happens after Health Coaching? Observational Study 1 Year Following a Randomized Controlled Trial." *The Annals of Family Medicine* 14(3): 200–207.

Song, B., K.-H. Wolf, M. Gietzelt, O. Al Scharaa, U. Tegtbur, R. Haux, and M. Marschollek. 2010. "Decision Support for Teletraining of COPD Patients." *Methods of Information in Medicine* 49(1): 96–102.

Totten, Annette M., Dana M. Womack, Karen B. Eden, Marian S. McDonagh, Jessica C. Griffin, Sara Grusing, and William R. Hersh. 2016. *Telehealth: Mapping the Evidence for Patient Outcomes from Systematic Reviews.* AHRQ Comparative Effectiveness Technical Briefs. Rockville, MD: Agency for Healthcare Research and Quality (US). http://www.ncbi.nlm.nih.gov/books/NBK379320/.

Tuckson, Reed V., Margo Edmunds, and Michael L Hodgkins. 2017. "Telehealth." *New England Journal of Medicine* 377(16): 1585–1592.

Welch, Garry, Andrew Balder, and Sofija Zagarins. 2015. "Telehealth Program for Type 2 Diabetes: Usability, Satisfaction, and Clinical Usefulness in an Urban Community Health Center." *Telemedicine Journal and E-Health* 21(5): 395–403. doi:10.1089/tmj.2014.0069.

Whittemore, Robin, and Jane Dixon. 2008. "Chronic Illness: The Process of Integration." *Journal of Clinical Nursing* 17(7b): 177–187.

Yom-Tov, Elad, Guy Feraru, Mark Kozdoba, Shie Mannor, Moshe Tennenholtz, and Irit Hochberg. 2017. "Encouraging Physical Activity in Patients with Diabetes: Intervention Using a Reinforcement Learning System." *Journal of Medical Internet Research* 19(10): e338. doi:10.2196/jmir.7994.

Zhang, Quan, Lingmei Ren, and Weisong Shi. 2013. "HONEY: A Multimodality Fall Detection and Telecare System." *Telemedicine and E-Health* 19(5): 415–429. doi:10.1089/tmj.2012.0109.

6 Social Factors in Human-Agent Teaming

Celso M. de Melo, Benjamin T. Files,
Kimberly A. Pollard, and Peter Khooshabeh

CONTENTS

6.1 SOCIAL FACTORS IN HUMAN-AGENT TEAMING

The last two decades have seen an explosion of interest in autonomous agents – such as robots, drones, self-driving cars, and home assistants – and the expectation is that these agents will become even more pervasive in society in the future (Bonnefon et al., 2016; de Melo et al., 2019; Stone & Lavine, 2014; Waldrop, 2015). As autonomous technology becomes integrated into our personal, social, and professional lives, humans will have to engage with it often and, in many cases, rely on it to accomplish their goals. In fact, human-agent teaming is expected to be critical to accomplishing the mission in increasingly complex and dynamic environments (Kott & Alberts, 2017; Kott & Stump, 2019). However, the success of these hybrid teams relies on a simple premise: humans will successfully collaborate with autonomous agents. But this premise should not be taken for granted. On the one hand, humans are very selective about with whom they cooperate, basing their decision on a multitude of factors including prior interaction, reputation, and shared group identity (Kollock, 1998; Rand & Nowak, 2013). On the other hand, many people may not trust autonomous technology, due to, for example, lack of experience and understanding of how it works (Gillis, 2017; Hancock et al., 2011; Lee & See, 2004). In this chapter, we review research supporting the argument that building machines that have appropriate social skills will encourage humans to treat them as social partners and, in turn, promote trust and cooperation in human-agent teams.

DOI: 10.1201/9781003215349-6

Humans are inherently social creatures. Our beliefs are influenced by our social context (Manstead & Fischer, 2001), we influence and are influenced by others (Van Kleef et al., 2010), we communicate with others to share information and synchronize our actions (Orbell et al., 1988), and we distinguish those that belong to our social groups from those that do not (Crisp & Hewstone, 2007; Tajfel & Turner, 1986). Moreover, we often anthropomorphize non-human others, and apply social heuristics learnt from interaction with other humans (Epley et al., 2007; Premack & Premack, 1995). Engaging in a social manner with non-human others supports intuitive explanations of others' behavior and brings familiar guidelines to unfamiliar situations (Reeves & Nass, 1996). For these reasons, several researchers argued for the development of social agents that, on the one hand, display cues that promote social engagement from humans and, on the other hand, simulate appropriate social behavior, including verbal and nonverbal communication and adapt behavior to the social context (Bates, 1994; Breazeal, 2003; Cassell, 2000; Gratch et al., 2002; Leite et al., 2013).

In this chapter, we review research on social factors that shape human-agent collaboration. We first look at the importance of natural language communication. Second, we emphasize the importance of nonverbal communication – in particular, emotion expression – to promote cooperation between humans and agents. Third, we review experimental studies indicating that humans readily apply social groups to agents, though tending to perceive agents, by default, as belonging to an out-group. Fourth, we look at how an individual's personality and traits shape their social interaction with agents. Finally, we discuss opportunities and challenges to the development of socially intelligent agents.

6.2 THEORETICAL FOUNDATIONS

Following a series of experimental studies showing that people treated machines in a social manner (Nass & Moon, 2000; Nass et al., 1996, 1997, 1999, 2000), Nass and colleagues advanced a general theory for human–machine interaction – the media equation theory (Reeves & Nass, 1996). According to this theory, people will intuitively treat machines in social settings as if they were social actors. The idea is that humans carry social heuristics learnt in human–human interaction to human–machine interaction automatically. A strict interpretation of the theory further argues that any social effect we see among humans could carry to human–machine interaction: "Findings and experimental methods from the social sciences can be applied directly to human-media interaction. It is possible to take a psychology research paper about how people respond to other people, replace the word 'human' with the word 'computer' and get the same results" (Reeves & Nass 1996, 28). Some of the studies supporting this view showed that people were polite to machines (Nass et al., 1999), formed positive impressions of machines perceived to be teammates (Nass et al., 1996), and applied social stereotypes to machines (Nass et al., 1997, 2000).

Blascovich et al. (2002), in contrast, proposed a more refined view – the social influence theory – which argues that machines are more likely to influence people, the higher the agency and realism of the machine. Agency increases with the perception that the machine is being controlled by a human; thus, an autonomous machine being controlled by algorithms would rank lower in this factor. Realism, or fidelity, relates to the photorealism of the machine (i.e., does it look like a human?), behavioral realism of the machine (i.e., does it behave like a human?), and the social realism of the machine (i.e., does it engage socially like a human?; Sinatra et al. 2021). According to this theory, it is possible to compensate for lack of agency by increasing realism. Studies in line with this view indicate that machines mirroring humans' nonverbal behavior can increase rapport (Gratch et al., 2007), simulating emotion expression in machines can increase cooperation (de Melo et al., 2014b), and photorealistic avatars that look like the user can increase compliance with an exercise regime (Fox & Bailenson, 2009).

Evidence from the emerging field of neuroeconomics presents further evidence that, even though humans can treat machines in a social manner, there are still important differences in the way humans behave with machines vs. humans. This research shows that people can reach different decisions and show different patterns of brain activation with machines in decision tasks, when compared to humans. Gallagher et al. (2002) showed that when people played the rock-paper-scissors game with a human there was activation of the medial prefrontal cortex, a region of the brain that had previously been implicated in mentalizing (i.e., inferring of other's beliefs, desires and intentions); however, no such activation occurred when people engaged with a machine that followed a known predefined algorithm. McCabe et al. (2001) found a similar pattern when people played the trust game with humans vs. machines, and others replicated this finding using prisoner's dilemma games (Kircher et al., 2009; Krach et al., 2008; Rilling et al., 2002). Sanfey et al. (2003) further showed that, when receiving unfair offers in the ultimatum game, people showed stronger activation of the bilateral anterior insula – a region associated with the experience of negative emotions – when engaging with humans, when compared to machines. de Melo et al. (2016) also showed that people made more favorable offers to humans than machines in various decision tasks and showed less guilt when exploiting machines. Overall, the evidence suggests that people experienced less emotion and spent less effort inferring mental states with machines than with humans. These findings are compatible with research showing that people perceive, by default, less mind in machines than in humans (Gray et al., 2007; Waytz et al., 2010). Denying mind to others or perceiving inferior mental ability in others, in turn, is known to lead to discrimination (Haslam, 2006) and can form the basis for out-group discrimination (Crisp & Hewstone, 2007; Tajfel & Turner, 1986), which could in some cases negatively impact human-agent team performance. In sum, even though there is increasing evidence that people are able to treat agents in a social manner, there is complementary evidence indicating that important differences remain. In the following sections, we describe mechanisms that support and encourage social interaction between humans and agents and, thus, help bring human-agent collaboration closer to what we see among humans.

6.3 VERBAL COMMUNICATION

One of the most fundamental ways in which humans communicate with one another is verbally via natural language. This makes natural language a promising modality for human–machine communication. Recent breakthroughs in cloud computing and machine learning techniques, combined with increased availability of large, well-annotated language corpora, have led to a dramatic uptick in research and development in the past decade, and in the widespread use of these technologies in everyday interactions. Natural Language Processing (NLP) technologies are more advanced – and more ubiquitous – than ever.

Perhaps the most familiar natural language enabled systems today are smart objects in the home (e.g., Alexa) and language-enabled assistants, such as Siri, on mobile devices. In our pockets, millions of us carry an NLP-enabled agent that we team with to do web searches, place orders, get directions, and even tell jokes. Siri, and similar technologies like Alexa, rely extensively on web-based data to answer users' questions and fulfill users' requests. However, such technologies are not well suited to addressing physical tasks or reasoning about items or processes that require sensing the physical world. Siri can look up directions to a sporting goods store where you can buy a basketball, but Siri can't find the basketball you already have stored in your closet.

Physically situated reasoning is a difficult challenge for autonomous and NLP-based systems, but progress is being made by uniting techniques (Gratch et al., 2015) used to develop non-physically situated conversational agents (e.g., Rizzo et al., 2011; DeVault et al., 2014; Traum et al., 2015) with robots and techniques to sense and navigate the real world (such as light-based ranging or image recognition technologies). One example is the Army Research Laboratory's JUDI (Joint Understanding and Dialogue Interface) project which involved developing a natural language-enabled search and navigation robot that can respond to spoken commands from a remotely located human teammate (Marge et al., 2016, Marge et al., 2017, Bonial et al., 2017; Lukin et al., 2018). The system can take action in the physical world and reply with text-based natural language confirmations and requests for clarification.

Even a remotely located robot with no visualization of facial features or intended emotional expressions can be perceived as a social partner simply by interacting with human users (Reeves & Nass, 1996). Humans interacting with versions of the JUDI system, for example, expressed interest in naming the robot, gave it encouraging verbal feedback, and inquired about the robot's gender (Henry et al., 2017; Pollard et al., 2018).

Creating a natural language dialogue system requires understanding what humans will want to say to the agent. Language-based systems can be constrained, involving a limited number of pre-determined commands and queries which the machine or agent is built to understand. However, a more flexible and intuitive system can be constructed to allow humans to speak to the agent using whatever phrasing they wish. Minimally constrained systems are more complex to build and often require extensive data collection to uncover, and account for, the wide diversity of verbal communication that humans may want to use with the system. This must be done so that the system can be built to accommodate this diversity. Humans can provide sample language during interviews or on questionnaires, or such data can be acquired by harvesting existing data in the wild such as from online chat language. A more direct and naturalistic way to collect data for some use cases is the Wizard of Oz method. This method encourages humans to interact with what they believe to be an autonomous agent. The human participant interacts with the agent as they would wish to, while a human researcher behind the scenes listens to (or reads) the language used and generates appropriate responses to act out through the "agent." The human researcher behind the scenes thus performs tasks such as speech recognition (if spoken language is to be used), natural language understanding, and natural language generation in the place of what will eventually become an automated system. This "wizard" may also execute other functions to eventually be performed by the autonomy (such as movement). A body of ecologically valid, application-specific human language samples, along with linked appropriate agent responses, can thus be collected before the envisioned autonomous system exists and can be used to then create the actual system. This method has been used fruitfully in the development of human-agent natural language dialogue systems for a variety of applications, such as counseling (e.g., DeVault et al., 2014), museum and history reenactments (e.g., Traum et al., 2015), and for physically situated navigation for robots (Bonial et al., 2017; Lukin et al., 2018).

Building machine systems to understand natural human utterances is just one piece of the puzzle. On the other side of the interaction, machines or agents built to produce verbal communication must express themselves in a manner that humans can understand and in ways that engender an appropriate level of rapport or trust in the agent. Here, we discuss two key examples of language use patterns that affect user response and performance outcomes in human-agent interactions, focusing specifically on agents used explicitly for social purposes, such as teaching.

Language style can influence a human's perception of, and response to, an autonomous agent. Teaching agents that use conversational language as opposed to formal language are often perceived more positively and lead to better human performance from the interaction (e.g., learning gains). Conversational language uses personalized pronouns (e.g., "you" and "me"), colloquial phrasing, and/or slang terms, which should engender a stronger feeling of social presence, and potentially greater motivation or interest, when interacting with the agent (the *personalization principle*, Moreno & Mayer, 2004). Some studies have found the use of conversational style language by pedagogical virtual agents to result in greater learning or transfer gains across a variety of academic subjects (Moreno & Mayer, 2000; Moreno & Mayer, 2004; Rey & Steib, 2013; Reichelt et al., 2014; Schneider et al., 2015a; Schrader et al., 2018). In some studies, conversational language features were associated with greater motivation (Kartal, 2010; Reichelt et al., 2014) and greater interest (Kartal, 2010). However, personalization can lead to worse performance in some cases (e.g., Kühl & Zander, 2017) or higher cognitive load (e.g., Kurt, 2011), and users' individual differences can play a role in how users respond to different language styles (Schrader et al., 2018).

While conversational language style can often be beneficial, research shows that it is also often beneficial for language to be polite. As with human–human verbal communication, agents that employ polite and face-saving (Brown & Levinson, 1987) language often yield improved human performance results. Polite, face-saving linguistic acts include the use of indirect wording or suggestions rather than direct wording or commands (e.g., *Could you turn the page?* or *For more information, you can turn the page.* vs. *Turn the page.*) The polite wordings help the receiver feel more as if they are exercising independent agency and personal control in completing the tasks.

This can be particularly important when agents are providing feedback regarding human performance (Mikheeva et al., 2019). The use of a polite language style was found to facilitate performance gains in a variety of learning domains (Wang et al., 2008; Schneider et al., 2015b), led to users choosing to spend more time engaging more with the learning material (Mikheeva et al., 2019), and was found to be more natural and less stress-inducing in a conversational interview context (Gebhard et al., 2014). However, users' individual differences can influence the effects of polite agent language on outcomes (e.g., Wang et al., 2008; McLaren et al., 2011), and non-direct language can be problematic in some settings, such as with healthcare robots (Lee et al., 2017). Additional research is needed to understand the various factors influencing human responses to agent politeness and conversational styles. The importance of individual differences and social group signifiers (e.g., accented speech, Khooshabeh et al., 2017) are discussed in separate sections of this chapter.

6.4 NONVERBAL COMMUNICATION

Complementing research on verbal communication, there has been growing interest on the role of nonverbal signaling in facilitating collaboration (Boone & Buck, 2003; de Melo et al., 2014; Gratch & de Melo, 2019; Lerner et al., 2015; Tickle-Degnen & Rosenthal, 1990; van Kleef & Côté, 2018; van Kleef et al., 2010). Rapport is emblematic and refers to a social phenomenon that occurs when people are highly engaged with each other, focused, mutually attentive, and enjoying the interaction (Tickle-Degnen & Rosenthal, 1990). Two important components in establishing rapport are nonverbal responsiveness – e.g., listening behaviors – and mimicry – for instance, mimicking the counterpart's posture. Establishing rapport has been shown to facilitate negotiation (Drolet & Morris, 2000), therapy (Tsui & Schultz, 1985), teaching (Fuchs, 1987), and caregiving (Burns, 1984), among others. Accordingly, human–computer interaction researchers have attempted to establish rapport between agents and humans, in particular, through nonverbal behavior (Bailenson & Yee, 2005; Gratch et al., 2006, 2007). Bailenson and Yee (2005) showed that mimicking agents were more persuasive and were rated more positively. Gratch et al. (2006, 2007) also showed that an agent that displayed listening behaviors – e.g., a nod in response to prosodic cues in the counterpart's speech – led to more fluent conversation and received more positive ratings.

One category of nonverbal signals, however, has received considerable attention due to its influence on human decision making and role in promoting cooperation: emotion expressions. In the last 20 years, there has been substantial experimental support for the interpersonal influence of emotion expressions in social decision making (for reviews, see Lerner et al., 2015; van Kleef & Côté, 2018; van Kleef et al., 2010), including effects on concession-making (van Kleef et al., 2004, 2006), emergence of cooperation (de Melo, Carnevale, Read, & Gratch, 2014b), fairness (Terada & Takeuchi, 2017; van Dijk et al., 2008), trust building (Krumhuber et al., 2007), and everyday life (Parkinson & Simons, 2009). Progress has also been made in understanding the pathways by which these effects operate. Broadly speaking, emotions can serve to evoke emotions in others via contagion (Lanzetta & Englis, 1989; Niedenthal et al., 2010) or can serve as information, revealing the experiencer's mental state (de Melo et al., 2014; Manstead & Fischer, 2001; van Kleef et al., 2010). This latter path is particularly interesting as it suggests a mechanism whereby people are able to "read other people's minds" by making appropriate inferences from other's emotion displays (de Melo et al., 2014b; Gratch & de Melo, 2019).

There is general agreement among emotion theorists that emotions are elicited by ongoing, conscious or nonconscious, appraisal of events with respect to the individual's beliefs and goals (Frijda, 1986; Scherer, 2001; Scherer & Moors, 2019). Different emotions result from different appraisals, as well as their associated patterns of physiological manifestation, action tendencies, and behavioral expressions. Expressions of emotions, therefore, reflect differentiated information about the expresser's appraisals and goals. Accordingly, researchers have noted that emotions serve important social functions, including communicating one's beliefs, desires, and intentions to others (Frijda & Mesquita, 1994; Keltner & Haidt, 1999; Keltner & Lerner, 2010; Morris & Keltner, 2000). In line with this view, Frank (1988, 2004) notes that emotion signals are ideal for identifying cooperators in society, especially since they tend to be harder to fake.

Several studies have now shown that emotion expressions can shape cooperation. de Melo and colleagues (de Melo et al., 2014; de Melo & Terada, 2019, 2020) revealed that emotion expressions compatible with an intention to cooperate (e.g., joy following mutual cooperation and regret following exploitation) led to increased cooperation in the iterated prisoner's dilemma. In contrast, emotion displays compatible with a competitive intention (e.g., joy following exploitation) hindered cooperation. These results emphasize the contextual nature of the effects of emotion, with the same exact expression leading to opposite effects according to the context in which it was shown. de Melo and Terada (2020) further showed that the effect of emotion expressions combine in interesting ways with the effect of actions: when actions were ambiguous or insufficient to convey the individual's intentions, the emotion signal became more relevant in the interaction; in contrast, when the counterpart's actions were clearly indicative of an intention to compete, emotion expressions had no effect. van Kleef and colleagues (van Kleef, 2016; van Kleef & Côté, 2018; van Kleef et al., 2010) further articulated the impact of emotion signals in more complex settings, such as negotiation. For instance, anger in negotiation led to increased concessions, as receivers inferred high aspirations on the sender's side (van Kleef et al., 2004). They further note several moderating factors – such as power and motivation to process information – on the effects of emotion expression on decision making (van Kleef, 2016).

Emotion expressions have also been shown to enhance human-agent interaction (Beale & Creed, 2009). In the context of the prisoner's dilemma, de Melo et al. (2009) showed that simulating facial expressions of emotion in an agent increased cooperation, when compared with an agent that showed no emotion. In follow-up work, de Melo et al. (2012) demonstrated that emotion expressions in agents also had the ability to hinder cooperation with humans, if they reflected competitive intentions (e.g., smile following exploitation). Moreover, several studies used experimental stimuli consisting of virtual faces – i.e., algorithms that simulate prototypical human facial expressions (de Melo, Carnevale, & Gratch, 2014a) – to research the social effects of emotions and, thus, arguably already provide support to the plausibility of simulating emotion in agents to promote collaboration with humans. Finally, researchers noted that emotion expressions can be used to overcome negative biases people have with agents (more on this in the next section) (de Melo & Terada, 2019; Terada & Takeuchi, 2017).

6.5 SOCIAL GROUPS

Humans often categorize others as belonging to distinct social groups, distinguishing "us" vs. "them," and this categorization influences collaboration as people are more likely to trust and cooperate with in-group than out-group members (Baillet, Wu, & De Dreu, 2014; Brewer, 1979; Crisp & Hewstone, 2007; Tajfel & Turner, 1986). Social identities, however, are complex and multifaceted. In many situations, more than one social category (e.g., gender, age, ethnicity) may be relevant. On the one hand, context can prime one category to become more dominant (or salient) and effectively exclude the influence of others. On the other hand, social categories can be simultaneously salient and have an additive effect on people's behavior (Crisp & Hewstone, 2007). These mechanisms based on multiple categories have, in fact, been proposed as the basis for reducing intergroup bias.

Experimental research indicates that people also engage in social categorization with agents. Nass et al. (1997) showed that participants perceived computers according to gender stereotypes, assigning more competence to computers with a female voice than a male voice on the topic of "love and relationships." Khooshabeh et al. (2017) showed that agents with voices with and without accent of the same culture, impacted perceptions of the appropriateness of the machine's decisions in social dilemmas. Researchers further showed that participants were more likely to cooperate (de Melo & Terada, 2019; de Melo, Carnevale, & Gratch, 2014a) and trust (Nass et al., 2000) agents that had virtual faces matching the participant's ethnicity.

However, as noted in the Theoretical Foundations section, people tend to make more favorable decisions with humans than agents, thus appearing to treat agents as out-group members by default. Accordingly, researchers looked at the possibility of associating positive social categories to compensate for negative social categories associated with agents. de Melo and Terada (2019) showed that conveying a cue for shared cultural identity in agents – through the ethnicity of the agent's virtual face – was sufficient to mitigate this bias. de Melo, Carnevale, and Gratch (2014a) further showed that creating a sense of belonging to the same team and sharing the same ethnicity could lead participants to be even more generous with agents than some humans. Complementary, de Melo and Terada (2019) showed that emotional expressions communicating affiliative intent (e.g., joy following mutual cooperation) could override initial expectations participants formed from social categorization.

6.6 INDIVIDUAL TRAITS

Individual differences influence social interactions. For example, a cross-cultural study of personality measured with the five-factor model found that Agreeableness and Conscientiousness were associated with quality of social interactions (Nezlek et al., 2011). Social Value Orientation is a relatively stable individual difference measure that describes the extent to which a person considers their own interests and the interests of others in their social interactions (Van Lange et al., 1997). Differences in social value orientation are associated with differences in negotiation behaviors (de Dreu & Van Lange, 1995). More rapidly changing individual differences, such as affective state, are associated with differences in social judgment (Forgas & Moylan, 1987). Although these examples come from studies of human social interaction, both the media equation theory (Reeves & Nass, 1996) and social influence theory (Blascovich et al., 2002), discussed earlier, suggest that individual differences that affect social interaction with humans should also affect social interactions with sufficiently realistic non-human social agents.

However, the extent to which findings from social psychology generalize to interactions with social agents might depend on the extent to which people treat agents as they would treat another person. The tendency to treat non-human animals and objects as human is called anthropomorphism (Epley et al., 2007), and individuals differ in the extent they tend to anthropomorphize. Differences in anthropomorphic tendency are associated with differences in the extent to which people interact with non-human agents as they would with a human (Waytz et al., 2010). Anthropomorphic tendency appears to be somewhat stable, but situational factors also affect anthropomorphic tendency. People who are lonelier anthropomorphize more, and being reminded of close, supportive social relationships is associated with less anthropomorphizing behaviors (Bartz et al., 2016). Similarly, Shin and Kim (2020) replicated the relationship between loneliness and increases in anthropomorphizing behavior, and they found that inducing feelings of loneliness with a writing task led to more anthropomorphizing behavior compared to control. More research is needed to understand how individual differences in anthropomorphizing mediate relationships between other characteristics and behavior toward social agents.

Studies of individual differences in social agent interaction can have practical utility, because social agents can potentially adapt to individual users' characteristics to increase the probability of some outcome the agent's designer prefers. Pedagogical agents (Sinatra et al., 2021) have been designed to leverage knowledge about an individual student to optimize that student's learning. As examples, PAL3 (Swartout et al., 2016) and GIFT (Sottilare et al., 2012) are systems that maintain a model of the student's learning based on the student's record of past activities, and they both can customize recommended activities based on a model of learning and forgetting to account for individual differences in background knowledge and learning speed. Both systems support delivering these recommendations via an onscreen agent. Other pedagogical agents have been designed to account for the learner's cultural knowledge in delivering feedback and content in cultural interaction training (Lane & Wray, 2012). Although some work has been done to examine the utility of stable personality traits in learner modeling, inferring the learner's characteristics from behavior might be a more promising approach (Abyaa et al., 2019).

Progress has been made in inferring human characteristics from behavior in the context of human/agent negotiation. For example, Sequeira and Marsella (2018) analyzed the offers humans made in the context of a structured human/agent negotiation task with the goal of discovering for each individual person a negotiating algorithm that reproduced the history of offers that person made. The characteristics of the best algorithms could be interpreted to summarize that person's approach to the negotiation. The human negotiators' social value orientation and Machiavellianism traits were also measured, and people with similar traits also had similar inferred algorithms. This illustrates an approach that could lead to negotiating social agents that learn a human partner's true negotiating style and objectives, potentially enabling the agent to engage in more effective negotiations.

In non-oppositional contexts, a social agent could request information about the human to enable it to effectively interact. The field of social psychology has developed many well-validated questionnaires and other instruments from which reliable individual differences measures can be calculated. For example, there are stable, population-level effects of message framing on decision outcomes (Tversky & Kahneman, 1981). However, an individual difference measure called regulatory focus (Higgins, 1998) accounts for variability in subjective (Higgins et al., 2003) and objective (Files et al., 2019; Glass et al., 2011) effects of message framing in the context of risk communication and performance feedback, among others. These findings suggest that social agents could more effectively communicate with individuals if they leverage knowledge of the individual's regulatory focus to frame messages about risk and opportunity.

In summary, individual differences affect several aspects of human social interaction, and the extent to which those same differences affect interactions with social agents might itself depend on individual differences in anthropomorphic tendency. Despite these complications, using individual differences to personalize social agent interactions could have big payoffs by creating a more effective experience. Progress has been made in identifying promising user characteristics on which to base agent customization, as has progress been made in inferring and measuring those characteristics. More research is needed to understand the complex interactions between user characteristics, social agent personalization, and interaction contexts to derive generalizable principles for effective personalized social agent interactions.

6.7   GENERAL DISCUSSION

We have argued that endowing agents with social skills can promote collaboration with humans. We reviewed literature indicating that humans regularly use these skills when working with others to achieve common goals. Moreover, we reviewed theory and experimental findings suggesting that humans will readily engage in a social manner with agents, especially the higher the social skill displayed by agents. The argument, therefore, suggests that designers cannot afford to ignore the broader social context in human-agent teams. Just as in human–human interaction, building rapport, trust, and cooperation with humans requires agents to skillfully navigate this social context, above and beyond the (non-social) functional aspects of the task.

We, thus, offer the following guidelines for designing socially intelligent agents:

- **Engage verbally and nonverbally with human teammates**: Communication between human and agent teammates can help ground the interaction, repair misunderstandings, and acknowledge mutual understanding through feedback cues. Nonverbal communication, especially through emotion expressions (de Melo et al., 2014), can be particularly important in communicating to humans an intention to cooperate, build trust, and mitigate mistakes (e.g., displays of regret).

- **Enhance perceptions of shared (social) group identity in agents**: Humans are likely to treat social agents as if they were out-group members, but this can be overcome by emphasizing shared social group membership (e.g., same ethnicity or accent as human teammates; Khooshabeh et al., 2017) or overriding negative social groups through clear signals of affiliative intent (e.g., appropriate expressions of emotion; de Melo & Terada, 2019).
- **Adapt the agent's behavior to the individual's personality and traits**: Agents that are able to customize their behavior to the individual's specific traits are more likely to collaborate successfully. Agents, therefore, should seek to learn about their human teammates through pre-interaction subjective personality scales, continuous physiological monitoring, and inferences from their actions. Once a model of the teammate is available, the agent can choose the optimal strategy – e.g., if a teammate is perceived to have a cooperative social value orientation, the agent can engage in cooperation from the start; however, if the teammate is perceived to have a competitive orientation, the agent can engage in tit-for-tat behavior to encourage cooperative behavior.

We, nevertheless, identify a few important open challenges to the successful adoption of socially intelligent agents:

- **Manage human teammates' expectations**: There is a long-standing debate in the robotics community about the so-called "uncanny valley" (Moris, 1970), which reflects a sudden shift in attitude, from empathy to revulsion, when the appearance of the robot begins to appear human-like while simultaneously failing to sufficiently act and look like a human. This uneasiness arises from the mismatch between the robot's appearance and the expectations that it generates in humans engaging with it. Similarly, with social agents it is critical that designers adjust the way social skill is expressed to the sophistication of the underlying model. For instance, it may be preferable to have a more robotic voice, rather than create the expectation that an agent is able to engage in open-ended conversation. This issue is particularly important as research suggests that people are more reluctant to trust autonomous machines given the lack of experience and understanding about how it works (Gillis, 2017). It is, therefore, essential to continuously manage the expectations in human teammates and help them form the correct mental model about the agents' (social) capabilities.
- **Ethical considerations**: The ability to exert social influence in humans is powerful, but potentially dangerous. Recently, there's been much discussion on the ethics of using technology that is able to perceive people's emotional states (Greene, 2019). There are concerns, on the one hand, about the accuracy of these algorithms and, on the other hand, about the appropriateness of engaging in emotion perception without the user's consent. Similarly, social agents will be able to make inferences about people's mental and affective states (e.g., is the individual trying to cooperate? Is the individual angry?) and engage in

behavior to change those states (e.g., express regret to reduce anger and promote cooperation). However, when is it appropriate to engage in this type of social manipulation, especially since often its effect is subconscious? Moreover, unlike humans, it is trivial for agents to fake social signals and, thus, steer humans' behavior toward its own interests. These are not easy issues, and it is beyond the scope of this chapter to solve them; however, it is important to acknowledge them, engage in cross-disciplinary debate, and encourage the ethical development of social agent technology from the start.

Socially intelligent agents promise to enhance the efficacy and efficiency of hybrid teams. They add a missing dimension to most current agent technology and bring human-agent collaboration closer to the kind of collaboration we see among humans. Moreover, because these agents can be designed from the ground up to optimally simulate social skill, if used ethically, they introduce a unique opportunity to create a more collaborative society.

6.8   AUTHOR NOTE

This research was supported by Mission Funds from the U.S. Army. The content does not necessarily reflect the position or the policy of any Government, and no official endorsement should be inferred.

Correspondence concerning this article should be addressed to Celso M. de Melo, CCDC U.S. Army Research Laboratory, Playa Vista, CA 90094, United States. Email: celso.m.de.melo.civ@mail.mil.

REFERENCES

Abyaa, A., Khalidi Idrissi, M., & Bennani, S. (2019). Learner modelling: Systematic review of the literature from the last 5 years. *Educational Technology Research and Development, 67*, 1105–1143.

Bailenson, J., & Yee, N. (2005). Digital Chameleons: Automatic assimilation of nonverbal gestures in immersive virtual environments. *Psychological Science, 16*, 814–819.

Baillet, D., Wu, J., & De Dreu, C. (2014). Ingroup favoritism in cooperation: A meta-analysis. *Psychological Bulletin, 140*, 1556–1581.

Bartz, J. A., Tchalova, K., & Fenerci, C. (2016). Reminders of social connection can attenuate anthropomorphism: A replication and extension of Epley, Akalis, Waytz, and Cacioppo (2008). *Psychological Science, 27*, 1644–1650.

Bates, J. (1994). The role of emotion in believable agents. *Communications of the ACM, 37*.

Beale, R., & Creed, C. (2009). Affective interaction: How emotional agents affect users. *Human-Computer Studies, 67*, 755–776.

Blascovich, J., Loomis, J., Beall, A., Swinth, K., Hoyt, C., & Bailenson, J. (2002). Immersive virtual environment technology as a methodological tool for social psychology. *Psychological Inquiry, 13*, 103–124.

Bonial, C., Marge, M., Artstein, R., Foots, A., Gervits, F., Hayes, C. J., Henry, C., Hill, S. G., Leuski, A., Lukin, S. M., Moolchandani, P., Pollard, K. A., Traum, D., & Voss, C. R. (2017). *Laying down the yellow brick road: Development of a wizard-of-oz interface for collecting human-robot dialogue. Proceedings of the Association for the Advancement of Artificial Intelligence Fall Symposium (AAAI 2017).*

Bonnefon, J.-F., Shariff, A., & Rahwan, I. (2016). The social dilemma of autonomous vehicles. *Science, 352,* 1573–1576.

Boone, R., & Buck, R. (2003). Emotional expressivity and trustworthiness: The role of nonverbal behavior in the evolution of cooperation. *Journal of Nonverbal Behavior, 27,* 163–182.

Breazeal, C. (2003). Toward sociable robots. *Robotic and Autonomous Systems,* 42, 167–175.

Brewer, M. (1979). In-group bias in the minimal intergroup situation: A cognitive-motivational analysis. *Psychological Bulletin, 86,* 307–324.

Brown, P., & Levinson, S. (1987). *Politeness: Some Universals in Language Usage.* New York, NY: Cambridge University Press.

Burns, M. (1984). Rapport and relationships: The basis of child care. *Journal of Child Care, 2,* 47–57.

Cassell, J. (2000). Embodied conversational interface agents. *Communications of the ACM, 43,* 70–78.

Crisp, R., & Hewstone, M. (2007). Multiple social categorization. *Advances in Experimental Social Psychology, 39,* 163–254.

de Dreu, C., & Van Lange, P. (1995). The impact of social value orientations on negotiator cognition and behavior. *Personality and Social Psychology Bulletin, 21,* 1178–1188.

de Melo, C., Carnevale, P., & Gratch, J. (2012). The impact of emotion displays in embodied agents on emergence of cooperation with people. *Presence: Teleoperators and Virtual Environments Journal, 20,* 449–465.

de Melo, C., Carnevale, P., & Gratch, J. (2014a). Using virtual confederates to research intergroup bias and conflict. In *Proceedings of Academy of Management (AoM).*

de Melo, C., Carnevale, P., Read, S., & Gratch, J. (2014b). Reading people's minds from emotion expressions in interdependent decision making. *Journal of Personality and Social Psychology, 106,* 73–88.

de Melo, C., Marsella, S., & Gratch, J. (2016). People do not feel guilty about exploiting machines. *ACM Transactions of Computer-Human Interaction, 23.* 10.1145/2890495.

de Melo, C., Marsella, S., & Gratch, J. (2019). Human cooperation when acting through autonomous machines. *Proceedings of the National Academy of Sciences U.S.A., 116,* 3482–3487.

de Melo, C., & Terada, K. (2019). Cooperation with autonomous machines through culture and emotion. *PLOS ONE,* https://doi.org/10.1371/journal.pone.0224758.

de Melo, C., & Terada, K. (2020). The interplay of emotion expressions and strategy in promoting cooperation in the iterated prisoner's dilemma. *Scientific Reports, 10,* 14959.

de Melo, C., Zheng, L., & Gratch, J. (2009). *Expression of moral emotions in cooperating agents.* In *Proceedings of the International Conference on Intelligent Virtual Agents.*

DeVault, D., Artstein, R., Benn, G., Dey, T., Fast, E., Gainer, A., Georgila, K., Gratch, J., Hartholt, A., Lhommet, M., Lucas, G., Marsella, S., Morbini, F., Nazarian, A., Sherer, S., Stratou, G., Suri, A., Traum, D., Wood, R., … Morency, L-P. (2014). *SimSensei Kiosk: A virtual human interviewer for healthcare decision support. Proceedings of the 2014 International Conference on Autonomous Agents and Multi-Agent Systems.*

Drolet, A., & Morris, M. (2000). Rapport in conflict resolution: accounting for how face-to-face contact fosters mutual cooperation in mixed-motive conflicts. *Experimental Social Psychology, 36,* 26–50.

Epley, N., Waytz, A., & Cacioppo, J. (2007). On seeing human: a three-factor theory of anthropomorphism. *Psychological Review, 114,* 864–886.

Files, B., Pollard, K., Oiknine, A., Passaro, A., & Khooshabeh, P. (2019). Prevention focus relates to performance on a loss-framed inhibitory control task. *Frontiers in Psychology,* 10. https://doi.org/10.3389/fpsyg.2019.00726

Forgas, J., & Moylan, S. (1987). After the movies: Transient mood and social judgments. *Personality and Social Psychology Bulletin, 13,* 467–477.

Fox, J., & Bailenson, J. (2009). Virtual self-modeling: The effects of vicarious reinforcement and identification on exercise behaviors. *Media Psychology, 12*, 1–25.

Frank, R. (1988). *Passions within reason: The Strategic Role of the Emotions*. New York: Norton.

Frank, R. (2004). Introducing moral emotions into models of rational choice. In A. Manstead, N. Frijda & A. Fischer (Eds.), *Feelings and Emotions* (pp. 422–440). New York, NY: Cambridge University Press.

Frijda, N. (1986). *The Emotions*. Cambridge, UK: Cambridge University Press.

Frijda, N., & Mesquita, B. (1994). The social roles and functions of emotions. In S. Kitayama & H. Markus (Eds.), *Emotion and culture: Empirical Studies of Mutual Influence* (pp. 51–87). Washington, DC: American Psychological Association.

Fuchs, D. (1987). Examiner familiarity effects on test performance: implications for training and practice. *Topics in Early Childhood Special Education, 7*, 90–104.

Gallagher, H., Anthony, J., Roepstorff, A., & Frith, C. (2002). Imaging the intentional stance in a competitive game. *NeuroImage, 16*, 814–821.

Gebhard, P., Baur, T., Damian, I., Mehlmann, G., Wagner, J., & André, E. (2014). Exploring Interaction Strategies for Virtual Characters to Induce Stress in Simulated Job Interviews. *Proceedings of the 2014 International Conference on Autonomous Agents and Multi-Agent Systems*, 661–668.

Gillis, J. (2017). Warfighter trust in autonomy. *DSIAC, 4*, 23–29.

Glass, B., Maddox, W., & Markman, A. (2011). Regulatory fit effects on stimulus identification. *Attention, Perception, & Psychophysics, 73*, 927–937.

Gratch, J., & de Melo, C. (2019). Inferring intentions from emotion expressions in social decision making. In U. Hess, & S. Hareli (Eds.), *The Social Nature of Emotion Expression* (pp. 141–160). Springer.

Gratch, J., Hill, S., Morency, L.-P., Pynadath, D., & Traum, D. (2015). Exploring the Implications of Virtual Human Research for Human-Robot Teams. In R. Shumaker & S. Lackey (Eds.), *Virtual, Augmented and Mixed Reality* (Vol. 9179, pp. 186–196). Springer.

Gratch, J., Okhmatovskaia, A., Lamothe, F., Marsella, S., Morales, M., Werf, R., & Morency, L.-P. (2006). *Virtual rapport*. In *Proceedings of 6th International Conference on Intelligent Virtual Agents*.

Gratch, J., Rickel, J., Andre, E., Badler, N., Cassell, J., & Petajan, E. (2002). Creating interactive virtual humans: Some assembly required. *IEEE Intelligent Systems*, 17, 54–63.

Gratch, J., Wang, N., Gerten, J., Fast, E., & Duffy, R. (2007). Creating rapport with virtual agents. In *Proceedings of Intelligent Virtual Agents (IVA'07)*. Springer Berlin Heidelberg, 125–138.

Gray, H., Gray, K., & Wegner, D. (2007). Dimensions of mind perception. *Science, 315*, 619.

Greene, G. (2019). The ethics of AI and emotional intelligence: Data sources, applications, and questions for evaluating ethics risk. *Partnership on AI Report*. Accessed from: https://www.partnershiponai.org/the-ethics-of-ai-and-emotional-intelligence/

Hancock, P., Billings, D., Schaefer, K., Chen, J., de Visser, E., & Parasuraman, R. (2011). A meta-analysis of factors affecting trust in human-robot interaction. *Human Factors, 53*, 517–527.

Haslam, N. (2006). Dehumanization: An integrative review. *Personality and Social Psychology Review, 10*, 252–264.

Henry, C., Moolchandani, P., Pollard, K. A., Bonial, C., Foots, A., Artstein, R., Hayes, C., Voss, C. R., Traum, D., & Marge, M. (2017). *Towards efficient human-robot dialogue collection: Moving Fido into the virtual world. Proceedings of ACL Workshop Women and Underrepresented Minorities in Natural Language Processing*.

Higgins, E. (1998). Promotion and prevention: Regulatory focus as a motivational principle. *Advances in Experimental Social Psychology, 30*, 1–46.

Higgins, E., Idson, L., Freitas, A., Spiegel, S., & Molden, D. (2003). Transfer of value from fit. *Journal of Personality and Social Psychology, 84*, 1140.

Kartal, G. (2010). Does language matter in multimedia learning? Personalization principle revisited. *Journal of Educational Psychology, 102*, 615–624.

Keltner, D., & Haidt, J. (1999). Social functions of emotions at four levels of analysis. *Cognition and Emotion, 13*, 505–521.

Keltner, D. & Lerner, J. (2010). Emotion. In D. Gilbert, S. Fiske & G. Lindzey (Eds.), *The Handbook of Social Psychology* (pp. 312–347). John Wiley & Sons.

Khooshabeh, P., Dehghani, M., Nazarian, A., & Gratch, J. (2017). The cultural influence model: When accented natural language spoken by virtual characters matters. *AI & Society, 32*, 9–16.

Kircher, T., Blümel, I., Marjoram. D., Lataster, T., Krabbendam, L., Weber, J., van Os, J., & Krach, S. (2009). Online mentalising investigated with functional MRI. *Neuroscience Letters, 454*, 176–181.

Kollock, P. (1998). Social dilemmas: The anatomy of cooperation. *Annual Review of Sociology, 24*, 183–214.

Kott, A., & Alberts, D. (2017). How do you command an Army of intelligent things? *Computer, 50*, 96–100.

Kott, A., & Stump, E. (2019). Intelligent autonomous things on the battlefield. In W. Lawlessm, R. Mittu, D. Sofge, I. Moskowitz, & S. Russell (Eds.), *Artificial Intelligence for the Internet of Everything (pp. 47–65)*. Academic Press.

Krach, S., Hegel, F., Wrede, B., Sagerer, G., Binkofski, F., & Kircher, T. (2008). Can machines think? Interaction and perspective taking with robots investigated via fMRI. *PLOS ONE, 3*, 1–11.

Krumhuber, E., Manstead, A., & Kappas, A. (2007). Facial dynamics as indicators of trustworthiness and cooperative behavior. *Emotion, 7*, 730–735.

Kühl, T., & Zander, S. (2017). An inverted personalization effect when learning with multimedia: The case of aversive content. *Computers & Education, 108*, 71–84.

Kurt, A. (2011). Personalization principle in multimedia learning: Conversational versus formal style in written word. *TOJET: The Turkish Online Journal of Educational Tecnology, 10*, 8.

Lane, H., & Wray, R. (2012). Individualized cultural and social skills learning with virtual humans. In A. M. Lesgold & P. J. Durlach (Eds.), *Adaptive Technologies for Training and Education* (pp. 204–221). New York, NY: Cambridge University Press.

Lanzetta, J., & Englis, B. (1989). Expectations of cooperation and competition and their effects on observer's vicarious emotional responses. *Journal of Personality and Social Psychology, 36*, 543–554.

Lee, D., & See, K. (2004). Trust in automation: Designing for appropriate reliance. *Human Factors, 46*, 50–80.

Lee, N., Kim, J., Kim, E., & Kwon, O. (2017). The influence of politeness behavior on user compliance with social robots in a healthcare service setting. *International Journal of Social Robotics, 9*, 727–743.

Leite, I., Martinho, C., & Paiva, A. (2013). Social robots for long-term interaction: A survey. *International Journal of Social Robotics, 5*, 291–308.

Lerner, J., Li, Y., Valdesolo, P., & Kassam, K. (2015). Emotion and decision making. *Annual Review of Psychology, 66*, 799–823.

Lukin, S. M., Gervits, F., Hayes, C. J., Moolchandani, P., Leuski, A., Rogers III, J. G., Sanchez Amaro, C., Marge, M., Voss, C. R., & Traum, D. (2018). *ScoutBot: A dialogue system for collaborative navigation. Proceedings of ACL 2018, System Demonstrations*, 93–98.

Manstead, A., & Fischer, A. (2001). Social appraisal: The social world as object of and influence on appraisal processes. In K. Scherer, A. Schorr, & T. Johnstone (Eds.), *Appraisal Processes in Emotion: Theory, Methods, Research* (pp. 221–232). Oxford, UK: Oxford University Press.

Marge, M., Bonial, C., Foots, A., Hayes, C., Henry, C., Pollard, K. A., Artstein, R., Voss, C. R., & Traum, D. (2017). *Exploring Variation of Natural Human Commands to a Robot in a Collaborative Navigation Task. Proceedings of the First Workshop on Language Grounding for Robotics*, 58–66.

Marge, M., Bonial, C., Pollard, K. A., Artstein, R., Byrne, B., Hill, S. G., Voss, C., & Traum, D. (2016). Assessing agreement in human-robot dialogue strategies: A tale of two wizards. In D. Traum, W. Swartout, P. Khooshabeh, S. Kopp, S. Scherer, & A. Leuski (Eds.), *Intelligent Virtual Agents* (pp. 484–488). Springer.

McCabe, K., Houser, D., Ryan, L., Smith, V., & Trouard, T. (2001). A functional imaging study of cooperation in two-person reciprocal exchange. *Proceedings of the National Academy of Science, 98*, 11832–11835.

McLaren, B. M., DeLeeuw, K. E., & Mayer, R. E. (2011). A politeness effect in learning with web-based intelligent tutors. *International Journal of Human-Computer Studies, 69*, 70–79.

Mikheeva, M., Schneider, S., Beege, M., & Rey, G. D. (2019). Boundary conditions of the politeness effect in online mathematical learning. *Computers in Human Behavior, 92*, 419–427.

Moreno, R., & Mayer, R. E. (2000). Engaging students in active learning: The case for personalized multimedia messages. *Journal of Educational Psychology, 92*, 724–733.

Moreno, R., & Mayer, R. E. (2004). Personalized messages that promote science learning in virtual environments. *Journal of Educational Psychology, 96*, 165–173.

Mori, M. (1970). The uncanny valley. *Energy, 7*, 33–35.

Morris, M., & Keltner, D. (2000). How emotions work: An analysis of the social functions of emotional expression in negotiations. *Research in Organizational Behavior, 22*, 1–50.

Nass, C., Fogg, B., & Moon, Y. (1996). Can computers be teammates? *International Journal of Human-Computer Studies, 45*, 669–678.

Nass, C., Isbister, K., & Lee, E.-J. (2000). Truth is beauty: Researching embodied conversational agents. In J. Cassell (Ed.), *Embodied Conversational Agents* (pp. 374–402). Cambridge, MA: MIT Press.

Nass, C., & Moon, Y. (2000). Machines and mindlessness: Social responses to computers. *Journal of Social Issues, 56*, 81–103.

Nass, C., Moon, Y., & Carney, P. (1999). Are people polite to computers? Responses to computer-based interviewing systems. *Journal of Applied Psychology, 29*, 1093–1110.

Nass, C., Moon, Y., & Green, N. (1997). Are computers gender-neutral? Gender stereotypic responses to computers. *Journal of Applied Social Psychology, 27*, 864–876.

Nezlek, J. B., Schütz, A., Schröder-Abé, M., & Smith, C. V. (2011). A cross-cultural study of relationships between daily social interaction and the five-factor model of personality. *Journal of Personality, 79*, 811–840.

Niedenthal, P., Mermillod, M., Maringer, M., & Hess, U. (2010). The Simulation of Smiles (SIMS) model: Embodied simulation and the meaning of facial expression. *Behavioral and Brain Sciences, 33*, 417–480.

Orbell, J., van de Kragt, A., & Dawes, R. (1988). Explaining discussion-induced cooperation. *Journal of Personality and Social Psychology, 54*, 811–819.

Parkinson, B., & Simons, G. (2009). Affecting others: Social appraisal and emotion contagion in everyday decision making. *Personality and Social Psychology Bulletin, 35*, 1071–1084.

Pollard, K. A., Lukin, S. M., Marge, M., Foots, A., & Hill, S. G. (2018). How we talk with robots: Eliciting minimally-constrained speech to build natural language interfaces and capabilities. *Proceedings of the Human Factors and Ergonomics Society Annual Meeting, 62*, 160–164.

Premack, D., & Premack, A. (1995). Origins of human social competence. In M. Gazzaniga (Ed.), *The Cognitive Neurosciences* (pp. 205–218). New York: Bradford.

Rand, D., & Nowak, M. (2013). Human cooperation. *Trends in Cognitive Science, 17*, 413–425.

Reeves, B., & Nass, C. (1996). *The Media Equation: How People Treat Computers, Television, and New Media Like Real People and Places*. New York, NY: Cambridge University Press.

Reichelt, M., Kämmerer, F., Niegemann, H. M., & Zander, S. (2014). Talk to me personally: Personalization of language style in computer-based learning. *Computers in Human Behavior, 35*, 199–210. https://doi.org/10.1016/j.chb.2014.03.005

Rey, G. D., & Steib, N. (2013). The personalization effect in multimedia learning: The influence of dialect. *Computers in Human Behavior, 29*, 2022–2028.

Rilling, J., Gutman, D., Zeh, T., Pagnoni, G., Berns, G., & Kilts, C. (2002). A neural basis for social cooperation. *Neuron, 35*, 395–405.

Rizzo, A., Sagae, K., Forbell, E., Kim, J., Lange, B., Buckwalter, J. G., Williams, J., Parsons, T. D., Kenny, P., Traum, D., Difede, J., & Rothbaum, B. O. (2011). *SimCoach: An intelligent virtual human system for providing healthcare information and support. Proc. Interservice/Industry Training, Simulation, and Education Conference (I/ITSEC) 2011*.

Sanfey, A., Rilling, J., Aronson, J., Nystrom, L., & Cohen, J. (2003). The neural basis of economic decision-making in the ultimatum game. *Science, 300*, 1755–1758.

Scherer, K. (2001). Appraisal considered as a process of multi-level sequential checking. In K. Scherer, A. Schorr & T. Johnstone (Eds.), *Appraisal Processes in Emotion: Theory, Methods, Research* (pp. 92–120). New York, NY: Oxford University Press.

Scherer, K., & Moors, A. (2019). The emotion process: Event appraisal and component differentiation. *Annual Review of Psychology, 70*, 719–745.

Schneider, S., Nebel, S., Pradel, S., & Rey, G. (2015a). Introducing the familiarity mechanism: A unified explanatory approach for the personalization effect and the examination of youth slang in multimedia learning. *Computers in Human Behavior, 43*, 129–138.

Schneider, S., Nebel, S., Pradel, S., & Rey, G. D. (2015b). Mind your Ps and Qs! How polite instructions affect learning with multimedia. *Computers in Human Behavior, 51*, 546–555.

Schrader, C., Reichelt, M., & Zander, S. (2018). The effect of the personalization principle on multimedia learning: The role of student individual interests as a predictor. *Educational Technology Research and Development, 66*, 1387–1397.

Sequeira, P., & Marsella, S. (2018). *Analyzing human negotiation using automated cognitive behavior analysis: The effect of personality. Proceedings of Cognitive Society Annual Meeting (CogSci)*, 1055–1060.

Shin, H., & Kim, J. (2020). My computer is more thoughtful than you: Loneliness, anthropomorphism and dehumanization. *Current Psychology, 39*, 445–453.

Sinatra, A., Pollard, K., Files, B., Oiknine, A., Ericson, M., & Khooshabeh, P. (2021). Social fidelity in virtual agents: Impacts on presence and learning. *Computers in Human Behavior, 114*, 106562. https://doi.org/10.1016/j.chb.2020.106562

Sottilare, R. A., Brawner, K. W., Goldberg, B. S., & Holden, H. K. (2012). The generalized intelligent framework for tutoring (GIFT). *Orlando, FL: US Army Research Laboratory– Human Research & Engineering Directorate (ARL-HRED)*.

Stone, R., & Lavine, M. (2014). The social life of robots. *Science, 346*, 178–179.

Swartout, W., Nye, B., Hartholt, A., Reilly, A., Graesser, A. C., VanLehn, K., Wetzel, J., Liewer, M., Morbini, F., Morgan, B., Wang, L., Benn, G., & Rosenberg, M. (2016). *Designing a personal assistant for life-long learning (PAL3). Proceedings of the 29th International Florida Artificial Intelligence Research Society Conference, FLAIRS 2016*, 491–496.

Tajfel, H., & Turner, J. (1986). The social identity theory of intergroup behavior. In S. Worchel & W. Austin (Eds.), *Psychology of Intergroup Relations* (pp 7–24). Chicago: Nelson-Hall.

Terada, K., & Takeuchi, C. (2017). Emotional expression in simple line drawings of a robot's face leads to higher offers in the ultimatum game. *Frontiers of Psychology, 8,* 724.

Tickle-Degnen, L., & Rosenthal, R. (1990). The nature of rapport and its nonverbal correlates. *Psychological Inquiry, 1,* 285–293.

Traum, D., Jones, A., Hays, K., Maio, H., Alexander, O., Artstein, R., Debevec, P., Gainer, A., Georgila, K., Haase, K., Jungblut, K., Leuski, A., Smith, S., & Swartout, W. (2015). New Dimensions in Testimony: Digitally Preserving a Holocaust Survivor's Interactive Storytelling. In H. Schoenau-Fog, L. E. Bruni, S. Louchart, & S. Baceviciute (Eds.), *Interactive Storytelling* (Vol. 9445, pp. 269–281). Springer.

Tsui, P., & Schultz, G. (1985). Failure of rapport: Why psychotheraputic engagement fails in the treatment of Asian clients. *American Journal of Orthopsychiatry, 55,* 561–569.

Tversky, A., & Kahneman, D. (1981). The framing of decisions and the psychology of choice. *Science, 211,* 453–458.

van Dijk, E., van Kleef, G. A., Steinel, W., & van Beest, I. (2008). A social functional approach to emotions in bargaining: When communicating anger pays and when it backfires. *Journal of Personality and Social Psychology, 94,* 600–614.

van Kleef, G. (2016). *The Interpersonal Dynamics of Emotion: Toward an Integrative Theory of Emotions as Social Information.* Cambridge, UK: Cambridge Univ. Press.

van Kleef, G. & Côté, S. (2018). Emotional dynamics in conflict and negotiation: Individual, dyadic, and group processes. *Annual Review of Organizational Psychology and Organizational Behavior, 5,* 437–464.

van Kleef, G., De Dreu, C., & Manstead, A. (2004). The interpersonal effects of anger and happiness in negotiations. *Journal of Personality and Social Psychology, 86,* 57–76.

van Kleef, G., De Dreu, C., & Manstead, A. (2006). Supplication and appeasement in negotiation: The interpersonal effects of disappointment, worry, guilt, and regret. *Journal of Personality and Social Psychology, 91,* 124–142.

van Kleef, G., De Dreu, C., & Manstead, A. (2010). An interpersonal approach to emotion in social decision making: The emotions as social information model. *Advances in Experimental Social Psychology, 42,* 45–96.

Van Lange, P., De Bruin, E., Otten, W., & Joireman, J. (1997). Development of prosocial, individualistic, and competitive orientations: Theory and preliminary evidence. *Journal of Personality and Social Psychology, 73,* 733–746.

Waldrop, M. (2015). No drivers required. *Nature, 518,* 20–23.

Wang, N., Johnson, W., Mayer, R., Rizzo, P., Shaw, E., & Collins, H. (2008). The politeness effect: Pedagogical agents and learning outcomes. *International Journal of Human-Computer Studies, 66,* 98–112.

Waytz, A., Gray, K., Epley, N., & Wegner, D. (2010). Causes and consequences of mind perception. *Trends in Cognitive Science, 14,* 383–388.

7 Human Elements in Machine Learning-Based Solutions to Cybersecurity

Mohd Anwar, Hilda Goins, Tianyang Zhang, and Xiangfeng Dai

CONTENTS

7.1 INTRODUCTION

Nowadays, we are witnessing an increase in the number of vulnerabilities and threats from web-based cyberattacks. Due to the rapid developments of global networking and communication technologies, a lot of our daily life activities such as jobs, education, business, and socialization are done in the cyberspace. The uncontrolled and anonymous infrastructure of the Internet presents an easy target for cyberattacks. These cyberattacks cause disruption of service and massive breach of data resulting in loss of privacy, economic loss, and other harms and inconveniences each year [1,2].

Cyberattacks are constantly on the rise and evolving in sophistication. By the time defenders identify attack patterns, the bad actors design newer attacks. In this arms race, some of the technical solutions to cybersecurity leverage various pattern recognition and anomaly detection methods. With the trove of data captured from monitoring of systems and networks, these cybersecurity mechanisms seek to employ machine

DOI: 10.1201/9781003215349-7

learning (ML) methods to discover latent attack patterns from the data. Humans play a vital role in the process of cybersecurity management and the use of the solutions depends on human factors.

In this chapter, we focus on three major cyberattacks: phishing, malware, and intrusion attempts. ML-based solutions have widely been proposed in the literature for the detection of these attacks. Some of these solutions report high accuracy of detection; however, the extent of deployment of these solutions in the production systems is unknown and reproducibility of these models is not studied. Although many of these solutions report high accuracy for detection of cyberattacks, often-times the efficacy of these models is limited to the data used to train them. Sometimes, these models are either not verified against publicly available gold-standard datasets or there is a lack of such datasets.

Because building of efficacious ML-based models and deployment of them on production systems significantly depend on human factors, we recognize the need to study human elements in ML-based solutions to cybersecurity. To employ supervised or semi-supervised machine learning methods, the data needs to be annotated by security experts and analysts. However, developing the labeled examples from previous attacks is expensive and time consuming. Also, because attackers constantly change their attack behavior, the updating of existing models is necessary and would require human involvement.

In this chapter, we discuss the challenges and limitations of ML-based solutions to cybersecurity in terms of their accuracy, reproducibility, deployability, and cost. We also explore human-in-the-loop to address the limitations of ML-based solutions. More specifically, we discuss strategies to tackle human labeling costs and challenges as well as to facilitate human integration in the end-to-end ML-based solution workflow.

7.2 ML-BASED SOLUTIONS FOR CYBERSECURITY

Machine learning methods are used to learn patterns of attacks by using past examples. These examples are driven from monitoring networks, endpoints, applications, and users. Three types of learning paradigms are utilized in devising solutions for cyberattacks: supervised, semi-supervised, and unsupervised. For supervised learning, detection models are developed from datasets consisting of features engineered from event logs along with labels, which indicate the presence or absence of an attack. The labeled dataset is split into training and test datasets, and models are trained on the training dataset. Afterward, a model's performance is evaluated against the test dataset. The model's efficacy is measured on how well the detection model classifies the attacks represented in the test dataset.

In the unsupervised learning paradigm, data points are grouped based on similarity or shared properties. Latent patterns are identified in unlabeled data. Since no label is given, this class of learning algorithms has the advantage of looking for attack patterns that have not been previously considered. Unsupervised learning can also be used to decrease the dimensionality of the input space prior to considering supervised learning. In semi-supervised learning, both labeled and unlabeled training data are used. Semi-supervised methods have attracted considerable interest in the machine learning community, showing significant use in applications such as text categorization and computer vision [3], which are also used in cybersecurity.

In addition to traditional machine learning (e.g., k-nearest neighbor, support vector machine, random forest, etc.), a new genre of machine learning called deep learning (DL) is increasingly implemented to devise cybersecurity solutions. These deep neural networks contain multiple processing layers and produce high-level flexible features from raw input data [4]. Convolutional Neural Networks (CNN) and Recurrent Neural Network (RNN) are two popular types of supervised DL architectures used in cybersecurity (e.g., [5]). The advantages of DL-based methods over traditional ML include automatic feature engineering and easier model updates.

7.2.1 PHISHING DETECTION

Phishing is a term coined in the mid-nineties describing a method by which hackers steal online account information. These lures are set out as hooks to "fish" for passwords and financial data from the "sea" of Internet users. Originally, phishing was limited to email scams where an email is sent to a victim from an entity giving them the impression that the communication is coming from a legitimate company or individual. However, the spread of phishing is no longer limited to traditional modalities such as email, SMS, etc. While mobile Internet and social networks have brought convenience to users, they have also been employed to spread phishing like QR code or spool phishing applications [6]. In addition, many phishing websites have similar design of the popular and legitimate sites, and end-users may not carefully check or search the official or original Uniform Resource Locators (URLs) and they have the potential to easily become a victim. Because the lifetime of a phishing attack is very short, there is a lack of long-term monitoring of the evolution and modification of phishing attacks [7], which poses a new detection challenge.

Machine learning-based methods are widely used in phishing website detection [8–11]. The detection of a phishing attack is a binary classification problem, for which training data must contain different features which should be strongly related to legitimate and phishing websites. Feature design and extraction are very important for developing detection models. For example, URL-based detection techniques analyze URL features to filter out suspicious malicious websites [12,13]. Though URL information can easily be acquired and extracted, attackers also have the flexibility to change or modify the URLs in a way that makes the features ineffective. Even though phishing pages usually maintain a similar visual appearance to the original pages, their contents are different. Thus, extracting web pages' content features to identify suspicious websites is another way for phishing detection [14–17]. Researchers developed techniques comparing the similarity of texts on a phishing website and the website targeted [14]. Other solutions compare the images of phishing and targeted pages to evaluate the visual similarity [16,17]. Another technique used is Cascading Style Sheets (CSS) of web pages as features to measure the similarity of suspicious and legitimate websites [18].

7.2.2 Malware Detection

Although the term "malware" has many meanings, it has generically been defined as any software whose objective is malevolent [19]. Malicious code is code that has been affixed, amended, replaced, or removed from software with the intent to sabotage the intended function of the software [20]. In malware (e.g., worms, trojans, viruses, ransomware, etc.) detection, static and/or dynamic analysis of the code is carried out. The static analysis examines the code structures and data properties. On the other hand, the execution behavior of the code is observed through code execution in dynamic analysis.

In the literature, various machine learning-based models report high accuracy for malware detection [21–25]. For example, through examining headers of executable files, researchers determined the metadata to be highly discriminative between malware and benign files that helped them identify the most discriminatory features. Furthermore, decision tree classifiers outperformed both logistic regression and Naive Bayes classifiers [21]. However, malware attacks continue to grow, and it is unclear if these models can be fully utilized in production environments to thwart the attacks. Specially, due to resource constraint nature of mobile devices, detection of mobile malware is a challenge [23,24]. It is unclear whether the results are reproducible and there is very limited effort to test these models "in the wild". As a result, these models are not always deployed in the production environment.

7.2.3 Intrusion Detection Systems (IDS)

To protect networks, intrusion detection is a widely used security mechanism. There are three different detection methods: misuse detection, anomaly detection, and hybrid detection. Misuse detection is used to detect known attacks using predefined patterns of various known attacks [26]. In this approach, abnormal system behavior is defined, whereas all other behaviors are considered normal. An attack is represented as a set of misuse patterns found in log files or data packets (header or data payload). Machine learning algorithms can detect misuse patterns that are learned from data collected by monitoring various systems/user activities. Once the model is built, user activity is presented to the model to determine any potential misuse [27]. In theory, misuse detection assumes that abnormal behavior can be defined by a simple model. However, although it is simple to add known attacks to the model, the disadvantage is the model's inability to recognize unknown attacks after the model has been trained without them.

For anomaly detection (i.e., recognizing deviations from normal patterns), prediction models are built using supervised learning methods (neural networks, support vector machine, k-nearest neighbors, etc.) which includes a labeled training set of normal and anomalous network and system activities [28]. Here, a dissimilarity detection technique is used to locate malicious behavior. Using normal activity as the benchmark, anything that deviates from this benchmark is considered anomalous. Normally, security experts create signatures which assist in the detection of known web-based threats. However, with the growth of the Internet-based computing, it has become a daunting task to keep up with these anomalous attacks and updating these signatures rapidly enough with the newly identified vulnerabilities.

In the hybrid intrusion detection model, both the misuse detection and anomaly detection are used [29–32]. In some of the literature, anomalous activities (e.g., network connections) are detected and then categorized using artificial neural network-based models. The need to include both misuse and anomaly detection has been justified because it raises the detection rate and lowers the false positive rate [30]. In a study on cloud-based architectures that combined architecture and knowledge-based intrusion detection (ABID and KBID), the use of a hybrid model reduced labeling time and enhanced detection and accuracy rates [31].

7.3 CHALLENGES AND LIMITATIONS OF ML-BASED SOLUTIONS

Although machine learning-based solutions have proven useful in thwarting cyberattacks, many challenges yet remain due to several limitations of ML-based solutions, which include the following:

- *Limited availability of appropriate data.* Supervised and semi-supervised learning require a large amount of annotated data – normal system data as well as anomalous and attack data – for model training and testing. Data annotation depends on human understanding of data. Thus, the training data becomes annotated correctly only if it can be understandable. The performance of a model relies heavily on the quality of the annotation. However, real-world data contains unstructured, noisy, dirty, and complex data that is not easy to be understandable [2]. The process of data annotation is time- and labor-intensive. For each instance of data annotation, there may be constraints on how much data one human expert can feasibly annotate in a unit of time. With the rapid development of information and communication technology, the amount of data has increased exponentially. It is a challenge to annotate the tremendous amount of data by human experts. Additionally, available experts or professionals are limited.
- *Class imbalance problem.* Class imbalanced datasets often occur in cybersecurity. Due to the rarity of the attacks, representation of the majority class, which holds the non-attack data far outnumbers the minority attack class. Consequently, there is a lot more data from normal systems vs. from systems under attacks. As a result, the algorithms may show overall high accuracy rates, hiding the real issue of low predictive accuracy for the minority class. Thus, it is imperative that the correct measure be used to appropriately represent the model performance.
- *Consequences of inaccurate output from ML models.* All machine learning-based detection systems will contain some amount of error. It is important to understand how these errors might affect the user experience. The user will need to decide which of the four measures [i.e., true positive (TP), false positive (FP), false negative (FN), true negative (TN)] of the confusion matrix more accurately represent the efficacy of the models and which can be discounted. If the model's emphasis is to accurately identify the attacks (positive cases), then one may assume that using TP (identifying the number of positive predictions of the attack class) is sufficient. However, this is not always the case [32].

Likewise, an inaccurately high FP rate could create unnecessary panic and may lead to complete mistrust in the system.

- *Zero-day attacks.* Zero-day attacks refer to being totally unaware of novel cyberattacks until they have occurred. With no prior knowledge, there is no time to thwart the attack. It is suspected that some software vendors simply hide actively exploited zero-days and report them as routine bug fixes [33]. According to data collected by Google's Project Zero security team, there were 11 zero-day vulnerabilities exploited in the first six months of the year 2020 and a total of 20 zero-days were recorded in 2019 [33]. Machine learning-based models require adequate time for building, testing, and deployment. Additionally, until the zero-data attack takes place, there is no knowledge about which vulnerability is being exploited in the attack. Because the bad actors are constantly changing their attack strategies, it is important for the model to be constantly updated to learn the attacks. If the detection model cannot be updated prospectively, the attacker can take advantage of yet-to-be-discovered vulnerabilities and evade detection. Thus, these models need to continuously adapt to de novo attacks. Any data-driven approach is as good as the dataset used to build the model.
- *Model trust and deployment.* The deployment of a model in the production system depends on the transparency of the model and confidence of human actors in the efficacy of the model. High detection accuracy on test data is not enough to trust a model. The human expert must be able to curate (understand and edit before using) the model. Thus, a significant amount of time and effort is required prior to the model's deployment. If the model is deemed insufficient, it may not be preferred, and the time and energy (including manpower and resources) used to build the model may have been wasted.

7.4   ADDRESSING THE CHALLENGES AND LIMITATIONS: HUMAN-IN-THE-LOOP MODEL BUILDING

The challenges and limitations of machine learning-based solutions can be mitigated with suitable human involvement in the process. Two types of human actors are critical for the successful development and deployment of ML-based solutions to cybersecurity: cybersecurity experts and ML experts. Traditionally, cybersecurity analysts were involved in monitoring and identification of the attacks patterns and taking corrective measures such as blocking these attacks. Also, machine learning engineers are knowledgeable in performing machine learning activities such as determining the best feature set for a learning model, creating the model by choosing the best features based on the problem at hand, and parameterizing the model. Collaboration of both types of experts is needed as it relates to the correct and timely response to cyberattacks. We are proposing that human elements should be integrated in the workflow so that the human can contribute, understand, build confidence, and improve the model. Human involvement in the cybersecurity solution workflow is described in the figure below.

In Figure 7.1, we present a five-layer human-ML collaboration cybersecurity model. As depicted in the figure, each layer of the model requires some level of human expert involvement. For the *cyber infrastructure* and *security data collection*

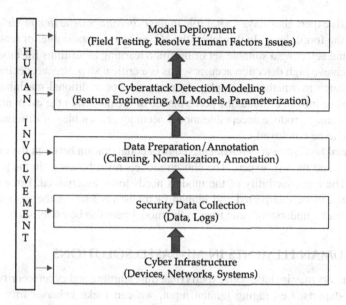

FIGURE 7.1 Human-ML collaboration model for cybersecurity.

layers, cybersecurity experts are heavily involved. For the *data preparation and annotation* layer, a collaboration between ML experts and cybersecurity experts is required. ML experts are also involved in *cyberattack detection modeling*. Finally, for an optimum *model deployment*, ML experts are again needed to work with cybersecurity experts.

The *cyber infrastructure* layer covers a diverse set of devices, systems, and networks. This layer includes personal computing, servers, IoT, clouds, Wi-Fi networks, WAN, LAN, etc. The cybersecurity expert is responsible for the instrumentation of monitoring tools and techniques for capturing activities in devices, systems, and networks.

The second layer, the *security data collection* layer, collects network data, database logs, and system logs in addition to application and user interaction data. In this layer, the cybersecurity analyst is responsible for ascertaining an unperturbed and protected collection of data as well as securing storage of data.

Once the data has been collected, the ML expert is responsible for preparing it for input to machine learning algorithms. Layer 3, the *data preparation and annotation* layer, is extremely critical for machine learning models. This task requires collaboration between cybersecurity experts and ML experts. For the ML expert, it is important to understand why and how a dataset was created or what preprocessing or cleaning had been done on the published data by the cybersecurity expert. Additional data cleaning as well as normalization of the data is required to preclude the effect of outliers, missing data, and calibration issues. Lastly, data annotation is achieved where labeling of malicious and benign data is verified. At this stage both the ML and cybersecurity experts need to work in collaboration to procure, clean, and understand the data as well as address any data inconsistencies and missing data issues.

The ML expert also plays a vital role in the *cyberattack detection modeling* layer, which is the fourth layer. In this layer, the ML expert is responsible for feature engineering and selecting a suitable set of machine learning algorithms for model building. To achieve high detection accuracy, this is a critical step. Various decisions must be made such as whether to employ deep learning or traditional machine learning paradigms or how to evaluate the performance of the model. In the event none of the algorithms may produce acceptable model accuracy, ensembles of different models may need to be employed.

The final layer, *model deployment*, requires collaboration between the cybersecurity expert and the ML expert. The chosen models must be tested in the production system. The reproducibility of the models needs to be ascertained. Error measurements need to be interpreted and human factor issues such as the working of the model must be understood, and trust in the model needs to be established.

7.5   HUMAN ELEMENTS IN ML-BASED SOLUTIONS

Human actors may include both experts (machine learning and cybersecurity experts) and non-experts. Leveraging human input, we can make cybersecurity solutions robust and deployable in production systems. Human involvement in the process will help build trust in these solutions. Here are some of the limitations of ML-based solutions that can be addressed through human involvement:

- *Labeling by crowd.* To solve the challenge of insufficient human experts, many studies propose solutions to annotate data by subsamples. There are some other studies that demonstrate examples of using non-experts in ML. One study investigated the annotation quality of non-experts and proposed an active selection criterion to query the best annotator from crowds for the most valuable instance [34]. A similar approach includes non-experts, where images are segmentized via super-pixels and non-experts only need to annotate the segments [35,36]. This approach reduces the annotation efforts and solves the challenge of insufficient experts as well.
- *Active learning.* In active learning, the learning algorithm can interactively query a user (called *teacher* or *oracle*) to label new data points with the desired outputs. Various algorithms are proposed to find the most informative and much smaller subset from the unlabeled data pool. Then the most informative subset is annotated by human experts, which means the experts do not need to annotate all the data. The approach achieves promising results with less human experts' efforts as in [37,38].
- *Interactive ML paradigm.* Interactive machine learning (iML) process iteratively allows human feedback in the training process to improve the accuracy of predictions [39]. For cyberattack detection models, it is important to obtain feedback from security analysts during the model-building process. This may also shorten the time taken for the algorithm to learn the attack patterns.

Figure 7.2 demonstrates an example of an iML process. First, a subset of data is annotated by domain experts. Then the annotated data is sent to train the model. Next, the intermediate results of the model and data are presented to the experts.

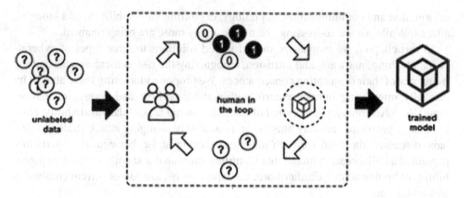

FIGURE 7.2 An example workflow of interactive machine learning.

They explore and draw insights about the data and the model and offer feedback in terms of new examples, features, and labels interactively. Afterward, the model is retrained with more annotated data. This iterative process is repeated until the performance of the model has adequately improved.

7.6   INTERPRETIVE MACHINE LEARNING

Machine learning models need to be understood not only by machine learning experts but also by domain experts through human-understandable explanation. Interpretability begins with data because understanding of the data leads to the selection of appropriate features and learning algorithms. Data interpretability is achieved through employing exploratory data analysis methods ranging from descriptive statistics to data visualization to principal component analysis (PCA) to clustering methods. Next, it is important to understand the model-building process. The domain experts should understand how the model works; this offers transparency of the model. The inner workings of the black-box models are not easy to interpret. Easy-to-interpret models are not as effective for recognizing complex patterns as are black-box models such as traditional and deep learning models.

Lastly, post hoc interpretability allows humans to interpret the model after it is trained and helps the domain expert analyze the strength and limitations of the model. It is helpful to interpret the black-box models by post hoc explainability. Once the model is trained, the ML expert analyzes the model by extracting information from the model [40]. In prediction-level interpretations, the focus is on model predictions based on the features that were used in the model prediction. On the other hand, dataset-level interpretation has a global focus, such as which patterns are associated with the predicted responses. With these two types of post hoc interpretations, a cybersecurity expert can understand the model's capabilities.

7.7   CONCLUSION

Over time, cyberattacks have become more frequent and devastating due to the increased dependence on the Internet for daily activities such as shopping and socializing. Attackers seem to be taking full advantage of our overdependence on information and communication technology, exploiting vulnerabilities, and lunching attacks. While we are addressing one attack, many more are being planned.

In this chapter, we emphasize more focused solutions to three types of cyberattacks: phishing, malware, and intrusion, recognizing the rise in these attacks and the frequency of their devastating consequences. We propose addressing these attacks by carefully integrating human experts within the workflow and leveraging human input. We have considered a myriad of issues including lack of data availability, class imbalance problem, zero-day attacks, and constantly changing attack strategies. We have discussed the need for both machine learning and cybersecurity experts and presented a collaboration model that identifies individual and collaborative responsibilities of human actors. Furthermore, we explore ways to address current challenges and limitations.

The black-box nature of state-of-the-art machine learning- or deep learning-based cyberattack detection models allows powerful prediction; however, the challenge is to produce explainable models while maintaining high performance. There is a trend in overemphasis in predictive power over the interpretability of the models. Nonetheless, cybersecurity experts need to trust a model for its deployment in production systems. Trust can be built from the involvement of ML experts and cybersecurity experts in the end-to-end process. It is critical to make the learned models as intelligible as possible for the complex task of cyberattack detection.

REFERENCES

[1] Phishing Activity Trends Reports. Accessed: February 1, 2021. https://apwg.org/trendsreports/

[2] Kaspersky Security Bulletin: Overall Statistics for 2017, Kaspersky Lab (2018).

[3] Zhu, X., Rogers, T., Qian, R., & Kalish, C. (2007, January). Humans perform semi-supervised classification too. *AAAI*, 2007, 864–870.

[4] LeCun, Y., Bengio, Y., & Hinton, G. (2015). Deep learning. *Nature*, *521*(7553), 436–444.

[5] Liu, H., Lang, B., Liu, M., & Yan, H. (2019). CNN and RNN based payload classification methods for attack detection. *Knowledge-Based Systems*, *163*, 332–341.

[6] Goel, D., & Jain, A. K. (2018). Mobile phishing attacks and defence mechanisms: State of art and open research challenges. *Computers & Security*, 73, 519–544.

[7] Cui, Q. (2019). Detection and Analysis of Phishing Attacks (Doctoral dissertation, University of Ottawa).

[8] Jain, A. K., & Gupta, B. B. (2016). A novel approach to protect against phishing attacks at client side using auto-updated white-list. *EURASIP Journal on Information Security*, 2016(1), 1–11.

[9] Zouina, M., & Outtaj, B. (2017). A novel lightweight URL phishing detection system using SVM and similarity index. Human-centric Computing and *Information Sciences*, 7(1), 1–13.

[10] Wang, Y., Agrawal, R., & Choi, B. Y. (2008). *Light weight anti-phishing with user whitelisting in a web browser*. In *2008 IEEE Region 5 Conference* (pp. 1–4). IEEE.

[11] Yang, P., Zhao, G., & Zeng, P. (2019). Phishing website detection based on multidimensional features driven by deep learning. *IEEE Access*, 7, 15196–15209.

[12] Almousa, M., & Anwar, M. (2019). *Detecting Exploit Websites Using Browser-based Predictive Analytics*. In *2019 17th International Conference on Privacy, Security and Trust (PST)* (pp. 1–3). IEEE.

[13] Ma, J., Saul, L. K., Savage, S., & Voelker, G. M. (2009). *Beyond blacklists: learning to detect malicious web sites from suspicious URLs.* In *Proceedings of the 15th ACM SIGKDD international conference on Knowledge discovery and data mining* (pp. 1245–1254).

[14] Zhang, Y., Hong, J. I., & Cranor, L. F. (2007). *Cantina: a content-based approach to detecting phishing web sites.* In *Proceedings of the 16th international conference on World Wide Web* (pp. 639–648).

[15] Cao, Y., Han, W., & Le, Y. (2008). *Anti-phishing based on automated individual white-list.* In *Proceedings of the 4th ACM workshop on Digital identity management* (pp. 51–60).

[16] Liu, W., Deng, X., Huang, G., & Fu, A. Y. (2006). An antiphishing strategy based on visual similarity assessment. *IEEE Internet Computing, 10*(2), 58–65.

[17] Fu, A. Y., Wenyin, L., & Deng, X. (2006). Detecting phishing web pages with visual similarity assessment based on earth mover's distance (EMD). *IEEE Transactions on Dependable and Secure Computing, 3*(4), 301–311.

[18] Mao, J., Tian, W., Li, P., Wei, T., & Liang, Z. (2017). Phishing-alarm: robust and efficient phishing detection via page component similarity. *IEEE Access,* 5, 17020–17030.

[19] Markel, Z., & Bilzor, M. (2014, October). *Building a machine learning classifier for malware detection.* In *2014 Second workshop on anti-malware testing research (WATeR)* (pp. 1–4). IEEE.

[20] Yerima, S. Y., Sezer, S., & Muttik, I. (2015). High accuracy android malware detection using ensemble learning. *IET Information Security, 9*(6), 313–320.

[21] Sahs, J., & Khan, L. (2012, August). *A machine learning approach to android malware detection.* In *2012 European Intelligence and Security Informatics Conference* (pp. 141–147). IEEE.

[22] Souri, A., & Hosseini, R. (2018). A state-of-the-art survey of malware detection approaches using data mining techniques. *Human-centric Computing and Information Sciences, 8*(1), 1–22.

[23] Gibert, D., Mateu, C., & Planes, J. (2020). The rise of machine learning for detection and classification of malware: Research developments, trends and challenges. *Journal of Network and Computer Applications, 153,* 102526.

[24] Christodorescu, M., Jha, S., Seshia, S. A., Song, D., & Bryant, R. E. (2005, May). *Semantics-aware malware detection.* In *2005 IEEE Symposium on Security and Privacy (S&P'05)* (pp. 32–46). IEEE.

[25] Dong-Her, S., Hsiu-Sen, C., Chun-Yuan, C., & Lin, B. (2004). Internet security: malicious e-mails detection and protection. *Industrial Management & Data Systems.*

[26] Depren, O., Topallar, M., Anarim, E., & Ciliz, M. K. (2005). An intelligent intrusion detection system (IDS) for anomaly and misuse detection in computer networks. *Expert systems with Applications, 29*(4), 713–722.

[27] Kemmerer, R. A., & Vigna, G. (2002). Intrusion detection: a brief history and overview. *Computer, 35*(4), supl 27–supl 30.

[28] Kamarudin, M. H., Maple, C., Watson, T., & Safa, N. S. (2017). A new unified intrusion anomaly detection in identifying unseen web attacks. *Security and Communication Networks, 2017,* 1–18.

[29] Meryem, A., & Ouahidi, B. E. (2020). Hybrid intrusion detection system using machine learning. *Network Security, 2020*(5), 8–19.

[30] Aydın, M. A., Zaim, A. H., & Ceylan, K. G. (2009). A hybrid intrusion detection system design for computer network security. *Computers & Electrical Engineering, 35*(3), 517–526.

[31] Rajendran, P. K., Muthukumar, B., & Nagarajan, G. (2015). Hybrid intrusion detection system for private cloud: a systematic approach. *Procedia Computer Science, 48,* 325–329.

[32] Subbarayalu, V., Surendiran, B., & Arun Raj Kumar, P. (2019). Hybrid network intrusion detection system for smart environments based on internet of things. *The Computer Journal*, *62*(12), 1822–1839.

[33] Cimpanu, C., "Zero Day" (2020), https://www.zdnet.com/article/google-eleven-zero=days-detected-in-the-wild-in-the-wild-in-the-first-half-of-2020, retrieved on 1/31/2021

[34] Li, S. Y., Jiang, Y., & Zhou, Z. H. (2015). Multi-label active learning from crowds. *arXiv* preprint arXiv:1508.00722.

[35] Rajchl, M., Lee, M. C., Oktay, O., Kamnitsas, K., Passerat-Palmbach, J., Bai, W., Damodaram, M., Rutherford, M.A., Hajnal, J.V., Kainz, B. & Rueckert, D. (2016). Deepcut: Object segmentation from bounding box annotations using convolutional neural networks. *IEEE transactions on medical imaging*, *36*(2), 674–683.

[36] Rajchl, M., Lee, M. C., Schrans, F., Davidson, A., Passerat-Palmbach, J., Tarroni, G., Alansary, A., Oktay, O., Kainz, B., & Rueckert, D. (2016). Learning under distributed weak supervision. *arXiv preprint arXiv:1606.01100*.

[37] Gal, Y., Islam, R., & Ghahramani, Z. (2017). *Deep bayesian active learning with image data*. In *International Conference on Machine Learning* (pp. 1183–1192). PMLR.

[38] Wu, J., Sheng, V. S., Zhang, J., Zhao, P., & Cui, Z. (2014). *Multi-label active learning for image classification*. In *2014 IEEE international conference on image processing (ICIP)* (pp. 5227–5231). IEEE.

[39] Amershi, S., Cakmak, M., Knox, W. B., & Kulesza, T. (2014). Power to the people: The role of humans in interactive machine learning. *Ai Magazine*, *35*(4), 105–120.

[40] Murdoch, W. J., Singh, C., Kumbier, K., Abbasi-Asl, R., & Yu, B. (2019). Definitions, methods, and applications in interpretable machine learning. *Proceedings of the National Academy of Sciences of the United States of America*, *116*(44), 22071–22080.

8 Cybersecurity in Smart and Intelligent Manufacturing Systems

Abbas Moallem

CONTENTS

DOI: 10.1201/9781003215349-8

8.1 INTRODUCTION

Manufacturing systems have gone through massive transformations in the last 40 years. The transformations can be categorized into four areas: a) geographical relocation; b) automation; c) supply chain management; and d) computerized equipment and machinery.

   a) Geographical Relocation: From early 1980, and due to advancements in real time global communications systems, the labor-intensive manufacturing sector gradually moved from industrialized countries to low labor wage countries. As these relocations took place, the manufacturing operations remaining used more automation to reduce labor costs. Soon after, the relocated manufacturing in low-wage countries also started to automate their manufacturing systems to achieve even lower costs.
   b) Automation: More and more industries started to automate the manufacturing process, and gradually, robotics and automation in high tech industries became the main trend.
   c) Supply Chain Management: Computerization of supply chain management and general production management with the Internet transformed manufacturing processes and made manufacturing more global.
   d) Computerized Equipment and Machinery: The equipment and machinery became computerized, smart, and intelligent, requiring interaction with sometimes complicated user interfaces and making them capable of working with the new digital age requirements.

This gradual change transformed production system components to operate in a network of interconnected system components: today these types of systems are referred to as Cyber-Physical Systems (CPS). According to the National Institute of Standards and Technology (NIST),

> Cyber-Physical Systems (CPS) comprise interacting digital, analog, physical, and human components engineered for function through integrated physics and logic. These systems will provide the foundation of the critical infrastructure, form the basis of emerging and future smart services, and improve our quality of life in many areas. Cyber-physical systems will bring advances in personalized health care, emergency response, traffic flow management.

In recent decades, progress in intelligent artificial, machine learning, cloud computing, and Internet of Things (IoT) have added more complexity to the manufacturing system and added two more attributes to manufacturing; the "Intelligent" and "Smart" system. There is not a common general agreement on the definition for these two terms.

According to NIST (Timothy et al. 2017), smart manufacturing systems are "fully-integrated, collaborative manufacturing systems that respond in real-time to meet changing demands and conditions in the factory, in the supply network, and in customer needs." The Smart Manufacturing Leadership Coalition (SMLC) definition states, "Smart Manufacturing is the ability to solve existing and future problems via an open infrastructure that allows solutions to be implemented at the speed of business while creating advantaged value" (Timothy et al. 2017).

8.2 INTELLIGENT AND SMART MANUFACTURING

The difference between "smart" and "intelligent" manufacturing is not very clear. Some consider that "intelligent" manufacturing optimizes production using various smart sensors, adaptive decision-making models, advanced materials, intelligent devices, and data analytics (Zhong Ray et al., 2017). Regardless, smart and intelligent manufacturing is impossible without the interconnection of smart machinery and IoT devices within a network of computers in a closed or open networking environment.

In the past, the security of manufacturing systems was focused only on the need to secure physical access to production sites. Nowadays, the manufacturing sites are not just closed entities; many components are in-network with globally dispersed manufacturing sites and outside partners. Hence, with smart and intelligent manufacturing systems, the risks of cyberattacks are exponentially growing every day, adding substantial financial impacts on enterprises. We now must recognize that another dimension added to such a network of computer connected machines to be the security of the system from cyberattackers who might want to access production information or interfere in some way with the system's regular operation.

8.2.1 MOTIVATION OF CYBERATTACKERS

The motivation for cyberattackers might be the same in manufacturing systems and non-manufacturing systems. However, several types of attacks are more specific to manufacturing systems. For example, manufacturing attackers' main motivations might be: a) stealing manufacturers' trade secrets and intellectual property; b) cyber-espionage; and c) sabotaging production systems and altering the design and process.

a) Manufacturers' Trade Secrets and intellectual property
   Manufacturing companies need to keep their products competitive in the global market and constantly strive to increase their competitiveness. Consequently, the demand for stolen intellectual property also increases. One of the main motivations of hackers targeting manufacturers is to obtain information regarding the planning, research, and development of a manufacturer's products. According to Verizon's 2018 Data Breach Investigations Report, around 47% of manufacturing breaches covered in the 2018 report involved intellectual property theft in gaining competitive advantage (Verizon, 2018).

b) Cyber-espionage
   Industrial espionage is another major threat to the manufacturing sector. According to the UK's GCHQ report cited by BBC (Ward, 2018), it is estimated that 34 separate nations have serious, well-funded cyber-espionage

teams. The threat from state-sponsored digital spies has been one of the most severe threats to networked global manufacturing systems.

c) Sabotaging Production System and Altering the Design and Process
Sabotage and alteration of an industrial system is often done by an insider or an employee. Such espionage is conducted for commercial purposes, national security, or cyber war.

The cases for sabotage and alteration of an industrial system are numerous. One of the most famous cases, "Stuxnet," a computer worm, targeted the programmable logic controllers (PLCs) to automate machine processes in Iranian Nuclear Facilities in 2010. The worm was designed to cause physical damage. Stuxnet seems to have destroyed about 1,000 (10%) of the centrifuges installed at the time of the attack. The malware was designed to attack two models of Siemens PLC (Siemens S7-125 and S7-41) (Zetter, 2014 & Fruhlinger, 2017).

In 2014, Remote Access Trojans (RAT) malware provided remote administrative control over an infected computer. To achieve this level of system access, a spear-phishing campaign that was used in watering hole attacks within the energy sector across the globe (Nelson, 2020) was employed. The sabotage attacks compromised industrial control systems (ICS), including supervisory control and data acquisition (SCADA) protocols, as well as the PLC, and Distributed Control Systems (DCS).

In January 2019, the press reported, based on Kaspersky findings, that an ASUS Live Update utility was compromised by cyber attackers. Usually, the Live Update utility comes pre-installed on the majority of ASUS computers sold, ensuring computer systems, such as drivers, apps, BIOS, and UEFI all receive upgrades and patches when they are due. This utility was compromised by cyberattackers. According to reports at the time,

> ASUS may have unwittingly spread malicious software to thousands of users through its update system. However, the cyberattackers appeared to be focused on a list of only 600 targets, hardcoded into the malware and identified by the unique MAC addresses used by their network adapters.
>
> **(Osborne, 2019)**

8.2.2 Manufacturing Cyber Vulnerability

Cybersecurity challenges in manufacturing systems are not much different than those of other systems. In general, cyber protection solutions developed for any purpose can also be applied to manufacturing systems. In doing so, typically, it is the manufacturing system's information and network components that will have some unique aspects that need to be considered. Below are the main particularities which must be reviewed.

8.2.2.1 Security Consideration at Design of the Systems

The security aspect of intelligent and smart manufacturing systems should be considered at the design phase. As such systems were coming online in recent years, many were designed without specialized network security in mind or with the assumption that the system will be isolated without considering the eventuality of global or even local cyberattacks. Therefore,

> Secure software development practices focusing on the prevention of software vulner-
> abilities, including the specification of security requirements, and the implementation of
> the security properties, including testing, code review, and patch management, have not
> been widely considered when building these systems.
>
> **(Tuptuk & Hailes, 2019)**

8.2.2.2 Heterogeneity of the Manufacturing Systems

Today, a smart manufacturing system domain includes many types of machinery, devices, and computers. The main machinery components have a long lifetime. The computerized parts of the equipment are not necessarily updated regularly due to the impact on systems automation interaction with the different physical processes (Tuptuk & Hailes, 2019).

8.2.2.3 Update and Upgrade

Speed, and sometimes immediacy in updating computer systems with security patches when vulnerabilities are discovered, is an essential security measure. However, in manufacturing systems implementing new patches seems to be slowed due to the impact that they might have on the overall system and eventual downtime that might be costly. Thus, some manufacturing systems might delay or avoid doing patch updates due to the risk associated with some of these challenges (Tuptuk & Hailes, 2019).

8.2.2.4 Performance versus Security

Many manufacturing system designs, to avoid operational downtime and other pro-ductivity challenges, focus more on operational efficiency and performance without sufficiently addressing evolving security concerns. Vulnerabilities also result from inadequate security policies and practices.

8.3 CYBERSECURITY IN MANUFACTURING SYSTEMS

The vulnerability of the manufacturing system from a cybersecurity perspective is not different than other systems. In this section, security issues are reviewed along with the related cases.

8.3.1 Encryption

Encryption is when the data is secured or encoded using some encryption keys, which are part of some algorithms. The secured or encoded data is unreadable with-out the encryption keys, and it looks like a meaningless language. In manufacturing systems, an important number of devices continuously communicate with each other and human agents. There are three types of data that needs to be encrypted:

- Data in motion: being transmitted over a network
- Data at rest: stored on machinery or IoT devices
- Data in use: in the process of being generated, updated, erased, or viewed

Data encryption is essential in communication between devices. However, encryption is not generally applied in manufacturing environments. Encryption key management techniques are often manual and require the operator to carry out the necessary actions such as renewing or revoking a key manually (Choi, Kim, Won, & Kim, 2009). ICS do not sufficiently encrypt data; consequently, hackers will have direct access to data that can be altered and modified, and disrupt processes once the network is breached.

8.3.2 AUTHENTICATION AND ACCESS MANAGEMENT

Authentication is a technique or mechanism to prove and validate an end-user or a computer program's identity. Industrial and manufacturing environments consist of hundreds or even thousands of devices. These devices use software and networking protocols to communicate with other devices and human operators. The ICS networks thus require a secure user authentication method. Lack of secure authentication and encryption security controls results in various types of attacks. Sometimes ICS attacks do not need to exploit software vulnerabilities. Once the network is breached, the attacker gains unfettered access to all the controllers and can alter their configuration, logic, and state to cause disruptions.

There are many cases reported due to authentication vulnerabilities. For example, Siemens has identified authentication vulnerabilities in the SIMATIC WinCC Sm@rt Client application. An attacker with local access could use these vulnerabilities to escalate privileges on the application or the servers with which the application communicates (Cybersecurity & Infrastructure Security Agency, 2015a). Another indicator was "Clorius Controls A/S ISC SCADA Insecure Java Client Web Authentication," The impacted Java client does not have a robust authentication mechanism. The client used an insufficient encoding to pass the credentials. Using weak credentials may allow those with access to the network to sniff traffic and decode credentials (Cybersecurity & Infrastructure Security Agency, 2015b).

In the case of "Accuenergy Acuvim II Authentication Vulnerabilities," the authentication bypass vulnerability allows access to the ethernet module web server interface settings without authenticating. The password bypass vulnerability enables an attacker to display passwords using JavaScript. A malicious user could create a denial of service for the webserver by changing the network settings (Cybersecurity & Infrastructure Security Agency, 2015).

8.3.3 OPERATOR ACCESS AUTHENTICATION

Since machinery constitutes the main component of manufacturing systems and interfaces also become more complex. The machines are becoming smarter and more intelligent with integrated computing capabilities as part of the cyber-physical system. Thus, accessing machinery interfaces is crucial for cyber protection and prevention of unauthorized accesses to the interfaces.

The first step for an attacker is to bypass to gain access to the human machine interface and control system. Consequently, the security of the authentication is fundamental to protect the system. Different authentication methods, including password authentication, one-time password authentication, and Kerberos authentication, are used to give access to legitimate users or operators. Also, the role-based access control model is used with time constraints in addition to the traditional access control model. Sometimes Analytical Hierarchy Process (AHP) is used to calculate each session's weight, according to which the time limit for each session was set. Often good practices, such as disabling unnecessary connections and changing default connection settings and secure passwords, are not implemented at production plants.

To improve verification of worker identity and operation-record visualization in workplaces, face recognition or voice recognition can be used for initial and on-demand identity verification of workers instead of a human agent verification (Okumura et al., 2019).

8.4 MACHINE TOOLS SECURITY

While machinery in manufacturing was becoming smarter and more intelligent, it also became vulnerable to cyberattacks. Manufacturing shop equipment is not isolated and is often connected to the Internet through wire or wireless connection. According to CYBERX, 84% of sites have at least one remotely accessible device and 40% of industrial sites have at least one direct connection to the Internet. While 53% of industrial sites have outdated Windows systems like XP, 69% have plain-text passwords traversing, and 57% of sites are still not running anti-virus protections that update electronic internal verification signatures automatically (CYBERX, 2019).

An aligned concern is that computerized numerical control machine tools can be hacked to gain access to a network and machine information. Unauthorized access might happen if the machinery is not secured. For example, the same password might be used for multiple logins or passwords may be used that are easy to guess. Also, and as mentioned above, there may be a lack of encryption. The hacker can then manipulate a machine's programming to make it create defective parts. In these cases, the operator might not even notice the changes.

8.4.1 INSIDER THREAT

As the term indicates, an insider threat is an attack coming from inside an organization, community, or group. The attacker(s), employees, or partners who have legitimate access to a system might deliberately or accidently misuse their privileges. The attackers' motivation might be to sabotage an organization, steal information or intellectual property (IP), conduct industrial espionage, or perpetrate fraud by modification, addition or deletion of an organization's data for personal gain. Manufacturing seems to be among the five industries with the highest percentage of insider threat incidents and privilege misuse. The average cost of insider threats is $8.86 million annually for a single manufacturing organization with more than 1,000 employees (McKee, 2019).

It is generally acknowledged that manufacturing companies are increasingly exposed to insider threats. According to a 2019 survey conducted by Fortinet, 68% of organizations feel moderately or extremely vulnerable to insider attacks, 68% of organizations confirm insider attacks are becoming more frequent, and 56% believe detecting insider attacks has become significantly to somewhat more problematic since migrating to the cloud. Finally, 62% think that privileged IT users pose the most significant insider security risk to organizations (FORTINET, 2019).

One of the recent insider threat cases happened at Tesla. According to Tesla CEO Elon Musk, a saboteur used his insider access to make "direct code changes to the Tesla Manufacturing Operating System under false usernames and exporting large amounts of highly sensitive Tesla data to unknown third parties" (Kolodny, 2018).

8.4.2 INTERNET OF THINGS (IoT)

Smart and intelligent manufacturing systems are composed of many interconnected devices with sensors. They communicate with and transmit data to each other through networks without a human to computer interaction, and with a multitude IoT sensors. The number of IoT devices is exponentially rising in all manufacturing systems. According to Gartner, 8.4 billion connected "Things" were in use in 2017, up 31% from 2016 [Gartner, 2017 & The Internet of Things (IoT), 2020]. These interconnected devices access a broad range of information about machinery and production systems. The progress in machine learning and artificial intelligence makes all these devices even more interactive and intelligent. Therefore, the need for all interconnected devices to be secure is even more important than before. If the devices are unsecured, someone can hijack them and cause severe damage.

IoT devices are vulnerable and can be hacked and utilized for larger, more sophisticated attacks. This means, of course, that malicious actors who target and hack IoT devices can use this access to gain control of personal information and even control a user's physical surroundings. There are numerous examples over the years of hackers exploiting security vulnerabilities in IoT devices. To prevent IoT devices from being hacked, it is important to have some essential security protocols implemented.

8.4.3 MALWARE

Malware is malicious software created or used to disrupt computer operation, gather sensitive information, or gain access to private computer systems. A virus can also be considered as malware, but viruses are technically different. Viruses are malicious code attached to a program that requires a user to start running it. Malware is itself an application working alone if installed on a system.

Worms are another kind of malware that start when one machine gets infected; the worm then spreads virally throughout the network on its own, infecting all the computers connected to it. Spyware malware spies on users and collects their private information, such as credit card numbers, without the user's knowledge. Another type of malware is referred to as a Trojan, which behaves like a regular program, yet creates a hidden door to breach the system's security. Hence, a Trojan is any malware that allows other malware to have easy access to the system. Lastly, ransomware encrypts all user information and files on the computer and asks for money to retrieve their files and data.

Since the early 2000s, malware has started to target critical infrastructures such as nuclear, pharmaceutical, aviation, and the electricity and water sectors (The Internet of Things (IoT), 2020). The most famous case is the previously mentioned "Stuxnet," a computer worm that targeted the PLCs used to automate machine processes in

Iranian Nuclear Facilities in 2010 (Zetter, 2014 & Fruhlinger, 2017 & Wolfgang Schwab, 2018).). In 2017, Triton malware was used to attack the petrochemical plant in the Kingdom of Saudi Arabia and caused it to shut down to prevent an explosion (Nicole & Clifford, 2018). BlackEnergy2 and Indutryoyer have attacked the electrical grid system in Ukraine (Osborne, 2018). Havex has exploited the CPS of different sectors in several European countries (Kovacs, 2015). The SolarWinds hack, one of the largest and most sophisticated attacks ever, happened in December 2020. Hackers secretly broke into SolarWinds systems and added malicious code into the company's Orion business software updates system. When SolarWinds, a major US information technology firm, sent out updates for bug fixes or new features to 33,000 customers, it included hacked code. Multiple US government agencies including the State Department, Treasury, and Homeland Security were targeted in the attack (FIREEYE, 2020; Satter et al., 2020).

8.4.4 PHISHING

Phishing is used to gain access to manufacturing systems and grows every year. The deception works by sending links to unsuspecting parties through junk email, instant messaging tools, mobile phones, short messages, or web pages. The links are promoted for clicks with false advertising claims, or with deceptive information from banks and other well-known institutions. Once the link is clicked, the malware is delivered to the system. Research has shown that even an experienced user can fail to distinguish between legitimate and illegitimate mail messages and embedded links and websites. Phishing attacks are already very sophisticated and are evolving every day. They are targeting everyone from individual users to large businesses and are causing billions of dollars of financial losses every year.

While the phishing model is evolving, its nature has not changed. Manufacturing, as any other type of industry, is also targeted heavily by phishing attacks to gain access to critical assets. According to Symantec, one in 384 emails sent to manufacturing employees contained malware and one out of 41 manufacturing employees was sent a phishing attack (Symantec, 2018). In the Ukrainian power grid attack (Verizon, (2018)), email phishing was the originating attack. In another case, an Austrian aerospace parts maker, a hacker, posed as the CEO and sent a phishing email to an entry-level accounting employee who transferred funds to an account for a fake project. In Leoni AG, a leading manufacturer of wire and cables, a finance employee in the company's Romania office was targeted by a phishing email claiming to be from the company's senior German executives.

8.5 ASSESSMENT AND PROTECTIVE TECHNOLOGIES

Manufacturing Cyber-Physical Systems (CPS) and physical control systems are not operating in isolation anymore. They are in a network of different interconnected systems that consist of a combination of various interconnected systems to monitor and manipulate real objects and processes. Today's manufacturing system is a network of embedded systems that interact with physical inputs and outputs. Since security was not always incorporated into manufacturing system design, they are more vulnerable to threats and security attacks. Besides, these systems are heterogeneous, since they use and operate a variety of IoT devices and components that communicate using different technologies and protocols, resulting in numerous vulnerabilities because of all the security gaps that can be exploited.

Consequently, a multi-layered approach is needed to prevent and mitigate the risk of cyberattack. The approach includes but is not limited to, vulnerability detection of the network, the platform, and systems security management.

Network vulnerabilities consist of detecting protective security weaknesses, such as open wired/wireless man-in-the-middle, eavesdropping, replay, sniffing, spoofing, and packet manipulation attacks (Amin et al., 2012 & Zhu et al., 2011). Platform vulnerabilities include hardware, software, configuration, and database vulnerabilities. Other vulnerabilities might be security management vulnerabilities due to lack of security guidelines, comprehensive security standards and guidelines, security policies, awareness training, and so on (Yaacoub, 2020).

The main security measures include Intrusion Detection Systems, Vulnerability Scanning (VS), Insider Attack Detection, Honeypots and Deception Techniques Detection, risk management, and penetration testing. Each is discussed in the following sections.

8.5.1 INTRUSION DETECTION SYSTEMS (IDS)

IDS's monitor operating system files, network traffic and events, and analyze the data gathered to detect signs of intrusion and detect other malicious activities. Upon detection of an untoward event, they alert the system administrators and, in some cases, initiate action. Since there are various ways that attackers might use to attack, IDS's can be an appliance or software. For example, attackers might try flooding or overloading the network, gathering data about the network, or finding the vulnerability to take control of the network. Consequently, intrusion detection tools often can prevent attackers from initially getting into the system.

8.5.2 VULNERABILITY SCANNING (VS)

VS is the technology that scans the network and computer system to check and identify known vulnerabilities. The system then generates risk exposure report. VS can be run on network devices such as firewalls, routers, switches, servers, and applications to find potential vulnerabilities. VS is conducted by automated tools that check if the networks, systems, and applications have security weaknesses that could expose them to attacks.

8.5.3 INSIDER ATTACKS DETECTION

Detecting insider threats is very challenging. The main challenges in effectively detecting these attacks include the fact the attackers have credentialed access to the network and services, use the applications with potential to leak data, or handle data that leave the protected boundary/perimeter, thus becoming subject to insider attack threats (Bruneau, 2020).

Insider threat detection technologies are monitoring and reporting tools. Technologies such as Data Loss Prevention (DLP) monitor abnormal access to sensitive data such as databases or information retrieval systems. Email security technologies monitor strange email exchanges, particularly outbound email. Privileged access management (PAM) monitors and controls people with higher privileged access to data, accounts, processes, systems, and IT environments. With advances in machine learning, user behavior and activity are analyzed and monitored. Thus, every unusual activity, exchange, or access is monitored and flagged.

8.5.4 Honeypots and Deception Techniques Detection

A honeypot is a computer system with applications and data that mimic the legitimate system and trick attackers into thinking it is a real system. Many different types of honeypots can be used to identify different types of threats. For example, a honeypot could be used in an inventory system or supply chain management system, or any other system commonly sought out by attackers (Cohen, 2006).

A honeypot is not a protection tool, but it is used to understand threats, capture how cybercriminals operate, and prevent future attacks. Various honeypots can be used based on the threat type. There is a variety of honeypot frameworks designed for the Manufacturing and CPS process. For example, the HoneyPhy aims to make it possible to construct honeypots for complex CPSs (Litchfield et al., 2020), or COCEAL is a framework to maximize key deception objectives, namely, concealability, detectability, and deterrence, while constraining the overall deployment cost (Bernieri et al., 2018).

8.5.5 Risk Management and the Cybersecurity Framework

The NIST proposes a framework to manage cybersecurity risk; identify, assess, and respond to risk; define risk tolerance; prioritize cybersecurity activities; and make informed cybersecurity decisions.

The NIST framework includes three parts; the Framework Core, the Framework Implementation Tiers, and the Framework Profiles. Each framework component reinforces the connection between business/mission drivers and cybersecurity activities.

> The Framework Core is a set of cybersecurity activities, desired outcomes, and applicable references that are common across critical infrastructure sectors...... Framework Implementation Tiers ('Tiers') provide context on how an organization views cybersecurity risk and the processes in place to manage that risk..... A Framework Profile ('Profile') represents the outcomes based on business needs that an organization has selected from the Framework Categories and Subcategories.
>
> (National Institute of Standards and Technology, 2018)

8.5.6 Penetration Testing of Manufacturing System

A penetration test (pen test) or ethical hacking is letting authorized experts try to hack a computer system to find out the system's vulnerability, helping find where the systems are vulnerable to then remove the areas of weakness.

Testing manufacturing systems to detect vulnerability is not as simple as the other systems. Manufacturing systems include an important number of older devices and machines without advanced features and newer machinery with sensors that evolve as the number of sophisticated cyberattacks increases. Consequently, testing manufacturing systems with diverse platforms and subsystems that are not necessarily centrally managed becomes more complex and challenging (Mahoney et al., 2017).

With the growing impact of cyberattacks in manufacturing systems, several different initiatives and research introduce a new generation of vulnerability assessment machine tools that are smarter and provide better protection and solutions. DeSmita et al. suggest a cyber-physical vulnerability impact analysis using decision trees. After completion of the intersection mapping, the decision trees are used to assess each node's vulnerability (DeSmita et al., 2017 & Chao Liu et al., 2018).

8.6 CONCLUSION

Manufacturing systems are continually changing due to rapid digitalization. Cyberattacks on manufacturing systems are increasing and lead to high costs and damage. Even though there is not much difference between manufacturing systems and other types of global systems (financial, health care, government, etc.) in terms of the type of attacks, the motivation of manufacturing system attackers differs from the motivations of other system intruders due to the value of cyber-espionage, trade secrets, and destructive or costly alterations of the manufacturing processes.

The cyber vulnerabilities of manufacturing systems also have some unique particularities. The particularities include diversified machinery and sensors, software updates, and access privilege monitoring. Additional vulnerabilities of manufacturing systems come with integrating a very diversified group of machines, IoT devices, and so on, as well as the lack of standardized communication protocols among the IoT devices that could lead to a more secure network.

We can see that cybersecurity technologies in the manufacturing area are changing fast, but so is the sophistication of the cyberattacks. Unfortunately, to date, there are not sufficient protective cyber technologies available to ensure that systems are protected from potential attacks.

A multi-layered approach needs to be implemented to prevent and mitigate the risk of cyberattacks so as to protect manufacturing systems. The approach needed would include, but is not limited to, vulnerability detection of the network and platform and enhanced risk management.

REFERENCES

Amin S., Litrico X., Sastry S., Bayen A.M. (2012). Cyber security of water SCADA systems-part I: Analysis and experimentation of stealthy deception attacks. *IEEE Trans. Control Syst. Technol.*; 21(5):1963–1970.

Bernieri G. et al. (2018). *A novel architecture for cyber-physical security in industrial control networks. IEEE 4th International Forum on Research and Technology for Society and Industry (RTSI) IEEE*; 2018. pp. 1–6.

Bruneau G. (2020). The history and evolution of intrusion detection. SANS Institute, Information Security Reading Room, 2020. https://www.sans.org/reading-room/white-papers/detection/history-evolution-intrusion-detection-344

Chao Liu C. et al. (2018). A systematic development method for cyber-physical machine tools. *Journal of Manufacturing Systems* 48 (2018) 13–24.

Chao Liu C. et al. (2018b). A systematic development method for cyber-physical machine tools. *Journal of Manufacturing Systems* 48 (2018) 13–24.

Choi D., Kim H., Won D., & Kim S. (2009). Advanced key-management architecture for secure SCADA communications. IEEE.

Cohen F. (2006). The use of deception techniques: honeypots and decoys. *Handbook of Information Security* 3(1): 646–655. https://www.semanticscholar.org/paper/The-Use-of-Deception-Techniques-%3A-Honeypots-and-Cohen/a41b7ab1c9bdad7cca82e739f9e08413eede7881?p2df

Cybersecurity & Infrastructure Security Agency (2015a). ICS Advisory (ICSA-15-013-01). Cybersecurity & Infrastructure Security, January 13, 2015. https://us-cert.cisa.gov/ics/advisories/ICSA-15-013-01

Cybersecurity & Infrastructure Security Agency (2015b). ICS Advisory (ICSA-15-013-02). Cybersecurity & Infrastructure, Security, October 13, 2014. https://us-cert.cisa.gov/ics/advisories/ICSA-14-275-02

CYBERX (2019). 2019 Global ICS & IIOT Report. CyberX, 2019. https://cdn2.hubspot.net/hubfs/2479124/CyberX%20Global%20ICS%20%2F%20IIoT%20Risk%20Report.pdf

DeSmita Z. et al. (2017). An approach to cyber-physical vulnerability assessment for intelligent manufacturing systems. *Journal of Manufacturing Systems* 43 (2017) 339–351. https://cra.org/ccc/wp-content/uploads/sites/2/2017/10/MForesight-Cybersecurity-Report.pdf

Egham, U.K., February 7, 2017. https://www.gartner.com/en/newsroom/press-releases/2017-02-07-gartner-says-8-billion-connected-things-will-be-in-use-in-2017-up-31-percent-from-2016

FIREEYE (2020). Highly evasive attacker leverages solarwinds supply chain to compromise multiple global victims with SUNBURST backdoor. Fireeye.com, December 13, 2020. https://www.fireeye.com/blog/threat-res

FORTINET (2019). Cyber security insider threat report. *FORTINET*, 2019. https://www.fortinet.com/content/dam/fortinet/assets/threat-reports/insider-threat-report.pdf

Fruhlinger J. (2017). What is Stuxnet, who created it and how does it work? Thanks to Stuxnet, we now live in a world where code can destroy machinery and stop (or start) a war. *cCSO*, August 22, 2017. https://www.csoonline.com/article/3218104/what-is-stuxnet-who-created-it-and-how-does-it-work.html

Gartner (2017). Gartner says 8.4 billion connected "things" will be in use in 2017, up 31 percent from 2016. Gartner.

Kolodny L. (2018). Elon Musk emails employees about 'extensive and damaging sabotage' by employee. *CNBC*, June 18 2018. https://www.cnbc.com/2018/06/18/elon-musk-email-employee-conducted-extensive-and-damaging-sabotage.html

Kovacs E. (2015). Attackers using Havex RAT against industrial control systems. *Security Week*, June 24, 2014, accessed November 10, 2015.

Litchfield S. et al. (2020). Rethinking the honeypot for cyber-physical systems. *IEEE Internet Computing* 20(5): 9–17. https://ieeexplore.ieee.org/abstract/document/7676152

McKee M. (2019). Insider threats: Manufacturing's silent scourge. *Industry Week*, April 25, 2019. https://www.industryweek.com/technology-and-iiot/article/22027503/insider-threats-anufacturings-silent-scourge

Mahoney T. et al. (2017). Cybersecurity for manufacturing: Securing the digitized and connected factory. MForesight, September 201, Report Number MF-TR-2017-0202

Nelson N. (2020). The impact of dragonfly malware on industrial control systems. SANS Institute 2020. https://www.sans.org/reading-room/whitepapers/ICS/impact-dragonfly-malware-industrial-control-systems-36672

Nicole P. & Clifford K. (2018). *A Cyberattack in Saudi Arabia Had a Deadly Goal*. Experts Fear Another Try. New York.

National Institute of Standards and Technology (2018). Framework for improving critical infrastructure cybersecurity. Draft version 1.1, April 16, 2018. https://nvlpubs.nist.gov/nistpubs/CSWP/NIST.CSWP.04162018.pdf

Okumura A. et al. (2019). Identity verification using face recognition for artificial-intelligence electronic forms with speech interaction. In A. Moallem (Ed.), *HCII 2019*, LNCS 11594, pp. 52–66. https://doi.org/10.1007/978-3-030-22351-9-4

Osborne C. (2018). Industroyer: An in-depth look at the culprit behind Ukraine's power grid blackout. *ZDNET*, April 30, 2018. https://www.zdnet.com/article/industroyer-an-in-depth-look-at-the-culprit-behind-ukraines-power-grid-blackout

Osborne C. (2019). Hijacked ASUS live update software installs backdoors on countless PCs worldwide. *ZDNET*, March 25, 2019. https://www.zdnet.com/article/supply-chain-attack-installs-backdoors-through-hijacked-asus-live-update-software/

Perlroth N. & Krauss C. (2018). A cyberattack in Saudi Arabia had a deadly goal. Experts fear another try. *Times*, March 15, 2018 November 1, 2018. https://www.nytimes.com/2018/03/15/technology/saudi-arabia-hackscyberattacks.html

Satter R., Bing CH., & Menn J. (2020). Hackers used SolarWinds' dominance against it in sprawling spy campaign. *Reuters*, December 15, 2020. https://www.reuters.com/article/idUSKBN28Q07P

Symantec (2018). 2018 internet security threat report. *ISTR*, Volume 23, 2018. https://docs.broadcom.com/doc/istr-23-2018-en

The Internet of Things (IoT). (2020). An overview. *Cong. Research Ser.*, Feb. 2020. https://crsreports.congress.gov/product/pdf/IF/IF11239

Timothy Z. et al. (2017). Cybersecurity framework manufacturing profile. National Institute of Standards and Technology, National Institute of Standards and Technology Internal Report 8183.

Tuptuk N. & Hailes S. (2019). Security of smart manufacturing systems. *Journal of Manufacturing System*, 47, 2018, pp. 93–106. https://discovery.ucl.ac.uk/id/eprint/10051762/1/1-s2.0-S0278612518300463-main.pdf

Verizon (2018). *Data Breach Investigations Report*, Verizon, 11th edition, 2018. https://enterprise.verizon.com/resources/reports/DBIR_2018_Report.pdf

Ward M. (2018). Staying one step ahead of the cyber-spies. *BBC News*, March 20, 2018. https://www.bbc.com/news/business-43259900

Wolfgang S. (2018). The state of industrial cybersecurity 2018. *Kaspersky*. https://ics.kaspersky.com/the-state-of-industrial-cybersecurity-2018/

Yaacoub J.P. (2020). Cyber-physical systems security: Limitations, issues, and future trends. *Microprocessors and Microsystems*, 77, September 2020, https://www.sciencedirect.com/science/article/pii/S0141933120303689?via%3Dihub

Zetter K. (2014). An unprecedented look at stuxnet, the world's first digital weapon. *WIRED*, December 3, 2015. https://www.wired.com/2014/11/countdown-to-zero-day-stuxnet/

Zhong Ray Y. et al. (2017). Intelligent manufacturing in the context of industry 4.0: A review. *Engineering* 3 (2017) 616–630.

Zhu B., Joseph A., & Sastry S. (2011). *A taxonomy of cyberattacks on SCADA systems*. In *International Conference on Internet of Things and 4th International Conference on Cyber, Physical and Social Computing*. IEEE, pp. 380–388. http://bnrg.cs.berkeley.edu/~adj/publications/paper-files/ZhuJosephSastry_SCADA_Attack_Taxonomy_FinalV.pdf

Index

Page numbers in *italics* denote figures; those in **bold** denote tables

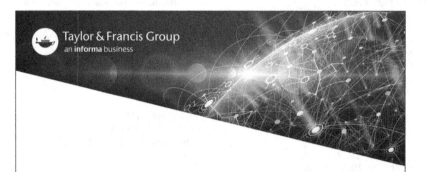

Printed in the United States
by Baker & Taylor Publisher Services

Printed in the United States
by Baker & Taylor Publisher Services